Five Step Process

Step 1: Conduct a Community Analysis
Step 2: Develop Community Partnerships
Step 3: Create an Intervention Strategy
Step 4: Implement
Step 5: Evaluate

U. S. Fire Administration

Public Fire Education Planning

A Five Step Process

FA-219/June 2008

FEMA

U.S. Fire Administration
Mission Statement

We provide National leadership to foster a solid foundation for local fire and emergency services for prevention, preparedness and response.

To the Reader:

Public Fire Education Planning: A Five-Step Process describes a systematic approach to designing, implementing, and evaluating community safety education programs.

This manual will help those of you who are new to community safety education understand some of the basic concepts about how to get started with an organization's community safety education programs. It also will assist you with some hints and techniques on a variety of topics, such as methods for locating partners to assist with community education or techniques for locating resources for your safety programs.

Acknowledgment:

The contributions of the following subject matter experts are gratefully acknowledged:

Management Systems
Training & Technology Co.
1331 Pennsylvania Ave., NW
Suite 1415
Washington, DC

Ed Kirtley, Chief,
City of Guymon Fire Department
Guymon, Oklahoma

Mike Weller
Life Safety Education
Hagerstown Fire Department
Hagerstown, Maryland

USFA Project Team
John Cochran,
Fire Management Specialist

Kathleen Gerstner,
Public Fire Education Specialist

Gerry Bassett,
Training Specialist

Table of Contents

Public Fire Education Planning: A Five-Step Process

Introduction

Conclusion
Bibliography
Appendix

Five-Step Process Summary

Step 1: Conduct a Community Risk Analysis

A community risk analysis is a process that identifies fire and life safety problems and the demographic characteristics of those at risk in a community.

Step 2: Develop Community Partnerships

A community partner is a person, group, or organization willing to join forces and address a community risk. The most effective risk reduction efforts are those that involve the community in the planning and solution process.

Step 3: Create an Intervention Strategy

An intervention strategy is the beginning of the detailed work necessary for the development of a successful fire or life safety risk reduction process. The most successful risk reduction efforts involve combined prevention interventions:

> **Education:** Providing information (facts) about risk and prevention.

> **Engineering:** Using technology to create safer products or modifying the environment where the risk is occurring.

> **Enforcement:** Rules that require the use of a safety initiative.

Step 4: Implement the Strategy

Implementing the strategy involves testing the interventions and then putting the plan into action in the community. It is essential that the implementation is well-coordinated and sequenced appropriately. Implementation occurs when the intervention strategy is put in place and the implementation plan schedules are followed.

Step 5: Evaluate the Results

The primary goal of the evaluation process is to demonstrate that the risk reduction efforts are reaching target populations, have the planned impact, and are demonstrably reducing loss. The evaluation plan measures performance on several levels, outcome, impact, and process objectives.

Introduction

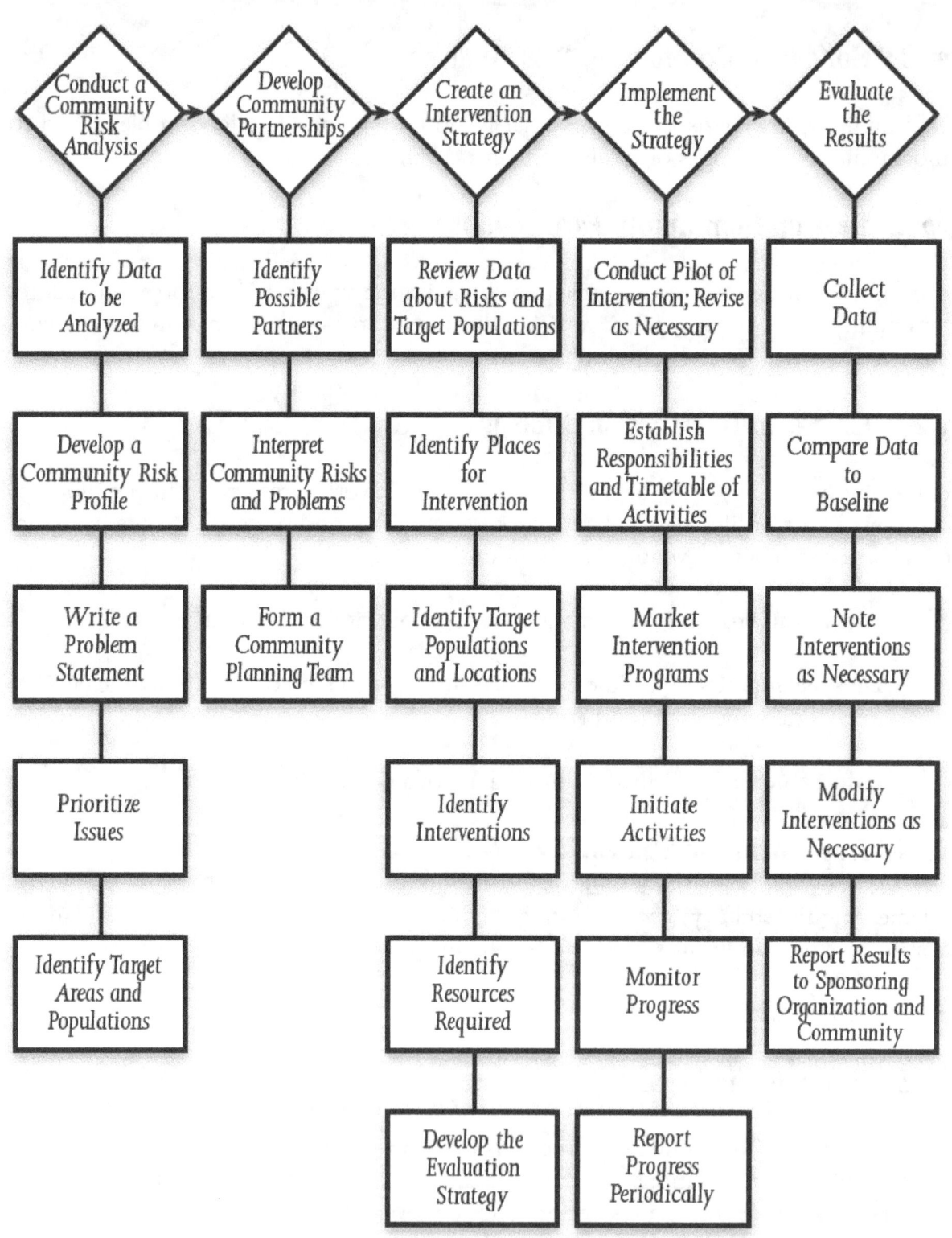

Conduct a Community Risk Analysis	Develop Community Partnerships	Create an Intervention Strategy	Implement the Strategy	Evaluate the Results
Identify Data to be Analyzed	Identify Possible Partners	Review Data about Risks and Target Populations	Conduct Pilot of Intervention; Revise as Necessary	Collect Data
Develop a Community Risk Profile	Interpret Community Risks and Problems	Identify Places for Intervention	Establish Responsibilities and Timetable of Activities	Compare Data to Baseline
Write a Problem Statement	Form a Community Planning Team	Identify Target Populations and Locations	Market Intervention Programs	Note Interventions as Necessary
Prioritize Issues		Identify Interventions	Initiate Activities	Modify Interventions as Necessary
Identify Target Areas and Populations		Identify Resources Required	Monitor Progress	Report Results to Sponsoring Organization and Community
		Develop the Evaluation Strategy	Report Progress Periodically	

Introduction

The purpose of this guide

Today fire departments use many terms for educational fire and injury prevention programs. Programs differ in size and approach, but the goals are the same: Change the behavior of the public so that there are fewer dangerous situations, fires, and injuries.

This guide uses a five-step planning process for developing and implementing successful fire and life safety public education programs. From identifying the fire and injury risks in the community, developing and implementing a program, and evaluating the results, planning is the process that ensures that the program strategies and initiatives really address the problems. This guide takes you step by step through that planning process.

A successful program follows a five-step process:

1.   **Conduct a community analysis.**

2.   **Develop community partnerships.**

3.   **Create an intervention strategy.**

4.   **Implement the strategy.**

5.   **Evaluate the results.**

The temptation to "just get something implemented" is hard to resist. Unfortunately, this is a trap. Yes, it's easy to schedule some presentations at a school; pass out brochures, stickers, and plastic helmets; and do some media interviews. But do those presentations address the community's worst fire or injury problems? Do the solutions being promoted really work? Is the appropriate target audience even being reached? Are community groups working together? Is the program being implemented in the best way?

A "ready, fire, aim" approach will not hit the target. It can give the impression that the department is out there educating the public, but may achieve little else. Successfully reducing fires and preventable injuries involves effective community planning. Notable public education programs around the country always prove this to be true.

This updated guide recognizes that fire prevention is now an important part of the larger goal of preventing injuries and unsafe situations. Fire departments often provide emergency medical services. Preventing injury, illness, and other unsafe situations is often part of the mission of saving lives.

What is the history of public education?

America Burning

Public education as an important component of a fire department's mission came to the forefront when *America Burning* was published in 1973. This report of the National Commission on Fire Prevention and Control, the forerunner of the United States Fire Administration (USFA), provided both insight into the scope of the American fire problem and possible solutions. *America Burning* identified the need for public education as part of an overall prevention strategy. "Public education about fire has been cited by many Commission witnesses and others as the single activity with the greatest potential for reducing losses."[1]

Motivational Psychology Fire Prevention Study

In 1974, Richard Strother published a report called "A Study of Motivational Psychology Related to Fire Prevention Behavior in Children and Adults," one of the first scientific studies to document the effectiveness of public education messages. It also demonstrated the importance of evaluation as an essential part of a public education program.

Public Fire Education Planning: A Five-Step Process

In 1977, the National Fire Prevention and Control Administration published a program for public educators detailing an effective planning model. *Public Fire Education Planning: A Five-Step Process* is a model that's been used for years by public educators to develop and implement fire safety education programs. The basic approach is still applicable today; however, changes in culture and the scope of problems require the updated planning model provided in this guide.

Learn Not to Burn

In 1979, the National Fire Protection Association (NFPA) introduced a new fire safety curriculum for use in elementary schools; *Learn Not to Burn*© was the first nationally available school-based curriculum to address fire safety behaviors for young children. The curriculum is based on research into the fire problem, the needs and characteristics of the target audience, and the behaviors appropriate for the target audience. In other words, *Learn Not to Burn*© is the culmination of a public education planning process.

Reaching the Hard to Reach

In 1981, the TriData Corporation, in cooperation with the USFA, published *Reaching the Hard to Reach*. This document emphasizes the need for local-level planning to identify the best approaches for reaching difficult-to-reach target audiences.

---

[1] National Commission on Fire Prevention and Control, *America Burning: The Report of the National Commission on Fire Prevention and Control* (Washington: U.S. Fire Administration, 1973) 105.

National Safe Kids Campaign

A new public education era began in 1988 when *SAFE KIDS Worldwide* was initiated in Washington, DC. The campaign is a cooperative effort between the National Children's Hospital and Johnson & Johnson Corporation. The goal is to reduce preventable childhood injuries, including those from fire. *Safe Kids* focuses on two key messages. First, public education initiatives must be based on local problems and target the people at risk. Each community must conduct its own planning process to identify specific problems and the best methods for addressing them. Secondly, any local initiative must involve the entire community. One organization alone can't make a big impact in preventing fires and injuries. Fire departments are recognizing that preventable injuries are as much of a problem in their communities as fires. Today, many fire departments have accepted the concept of working with other organizations, and have become key members of local *Safe Kids* coalitions.

Risk Watch

In 1995, the NFPA began working with the *National Safe Kids Campaign* to develop an all-risk injury prevention curriculum called *Risk Watch®*. Again, effective planning proved to be a key to the project's success. Using the planning process, the NFPA was able to identify the major risk areas for children, determine appropriate messages and behaviors, and develop successful methods for teaching the behaviors and messages. *Risk Watch®* is now an effective public education tool.

Lessons Learned

Since 1973, there have been many lessons learned about what constitutes effective public education initiatives at the local level. Experience shows that successful programs have the following characteristics:

- **There is strong individual and organizational commitment to the public education initiatives.** The initiatives involve more than a few presentations and distribution of brochures. Successful public educators understand the amount of effort involved and commit themselves to a long-term approach. In addition, the organization's leaders understand that the program requires full organizational commitment that includes time for personnel to work on the program, resources required to buy materials and equipment, and most importantly, visible program support from all department members and other local decisionmakers.

- **The program is based on a comprehensive planning process that identifies community fire and injury problems, and the people most likely to be involved.** Few, if any, successful programs are haphazard. Organizations with successful programs have used a planning process. In some cases this takes time. In other cases (for example, in a small community with a specific problem) the process may be shorter. The complete planning process should be used by organizations of all sizes.

- **Partnerships are established so that the community as a whole is involved in the solutions.** Solutions to fire and injury problems can be complex. Several local organizations should form a team and work together to solve community safety problems. This approach brings a variety of resources to bear on the problem and reduces the cost to any one organization. Everyone has a vested interest in the success of the program and in improving the quality of life of the target audience.

- **There is an evaluation of the program's results and processes.** Evaluation is crucial to determining if a program's goals have been achieved, and essential to planning future programs. Evaluation identifies what worked, what didn't work, and what to do in the future to be successful. The evaluation also provides the ammunition needed to market the program to the community.

What has improved?

Public education, along with engineering and enforcement (the three E's), have reduced fires and related deaths, burn injuries, and other types of preventable injuries. Consider the following fire and injury statistics:

- From 1995 to 2004, overall fire incidents declined steadily by 20 percent.

- The number of annual deaths due to fire declined 21 percent from 1995 to 2004. Now the annual average is fewer than 4,000 deaths each year.

- From 1995 to 2004, the number of injuries from residential fires decreased by 29 percent.

- The death rate from fire among children 14 and under declined by 68 percent from 1987 to 2005.

- The overall unintentional injury-related death rate among children 14 and under decreased by 45 percent between 1987 and 2005.

However, additional statistics indicate that more work is needed. Consider some different statistics:

- The greatest numbers of fire deaths occur in the home, with the majority in one- and two-family dwellings.

- In 2005, more than 1,400 child occupants (0-14) died in motor vehicle crashes and nearly half were unrestrained. In the same year, 203,000 child occupants were injured.

- Every year more than 5,000 American children ages 14 and under die from unintentional injury

- The elderly, the young, and the poor continue to be at a significantly higher level of risk from fire than the average person.

- While African-Americans account for 13 percent of the population, they account for 24 percent of the annual fire deaths.

- The amount of annual property loss due to fire each year averages $11 billion.

What needs to be done in the future?

The 2000 USFA report, *America Burning Recommissioned*, states, "There is wide acknowledgement and acceptance that public education programs on fire prevention are effective...no prevention effort can succeed without a public education component."[2]

Fire service leaders and public educators responsible for programs must use new educational approaches, methods, and processes, such as the following:

- **Greater organizational focus on prevention.** Fire departments must continue to increase the emphasis on prevention initiatives. Prevention is an important, cost-effective risk reduction tool. Fire chiefs, fire marshals, public educators, and all fire service leaders must be advocates for these initiatives. They should be the national leaders in fire and injury prevention efforts.

- **Improved data collection and analysis.** This directly supports the need for better planning. Currently, information on fires is available in most States through the National Fire Incident Reporting System (NFIRS). However, there are voids in available information. It is essential that each community gather data on fires and preventable injuries. This requires collaboration with agencies such as law enforcement, hospitals, burn units, health departments, and State fire marshals. Data analysis is part of the overall planning process, as well as the development of an effective public education program.

- **Integrated use of prevention interventions.** The most effective prevention programs incorporate education, code enforcement, and engineering interventions in one comprehensive prevention strategy. The three interventions working together hit the problem from all sides so the ability to reduce deaths and injuries is improved dramatically.

- **Improved technology.** Safety technology is progressing rapidly. No longer are the smoke alarm and automatic sprinkler system the only prevention tools available to the public educator. New technologies include digital projectors, fire detection systems that provide building occupants with more time to escape from a fire, and computers that increase the ability of the public educator to reach others with prevention messages. These technologies must be part of the public educator's toolbox and be integrated into education and prevention programs.

---

[2] U.S. Fire Administration, *America Burning Recommissioned* (Washington: Author, 2000) 24.

- **Higher level of prevention and public education training for emergency services personnel.** Firefighters and emergency medical technicians (EMTs) are routinely delivering public education presentations and assisting with public education activities. They must be adequately trained in educational methods. NFPA 1001, *Standard for Fire Fighter Professional Qualification* requires new firefighters to have the skills to deliver a public education presentation from a prepared lesson plan. NFPA 1021, *Standard for Fire Officers*, specifies the requirement of the ability to deliver presentations and also to develop a department public education program. These skills can be included in job descriptions and specified as a prerequisite for promotion. Every emergency service provider is a public educator at some level, and can benefit from public education training.

- **Changing the current paradigms about the causes of fire and injuries.** The term "accident" has been used for years to describe the cause of fires and injuries. By definition, an accident is an uncontrolled event—something that cannot be predicted or prevented. When matches are left in the reach of a young child and that child starts a fire with them, it is not an accident. It is an incident that could have been prevented. When a fire occurs in a home without a working smoke alarm and the residents are killed, it is not an accident. A working smoke alarm and education on home escape plans can prevent these deaths. The public educator must teach the public that injuries and fires are seldom accidents. Rather, they are predictable events that can be prevented through education, enforcement, and engineering initiatives.

Can I make a difference?

This is a question that many newly-appointed public educators ask. The answer is, without a doubt, yes! Public education has proved itself over and over as an effective prevention strategy. The *SAFE KIDS Worldwide* campaign estimates that, on average, a $33 smoke alarm generates $940 in benefits to society, such as medical costs and other costs associated with fire suppression and property loss. What a return on an investment of one dollar! *Safe Kids* also estimates that as many as 90 percent of unintentional injuries to children can be prevented.

Community risk reduction uses prevention processes to reduce or eliminate hazards and risks in the community, thus reducing the frequency and severity of fires and injuries. This effort requires planning. But the foundation for success is laid long before the planning step. First, it's important to have proactive individual and organizational attitudes about the community education program, and a strong personal and organizational commitment to making the program achieve its goals and objectives.

The community educator is the heart and the soul of the program. This person must motivate others to be involved and support the program. A sound, rational argument for your community education program is important. However, a positive attitude and strong commitment is what actually convinces others that a program should be implemented.

Department and community members will assess the level of commitment on the part of the department leadership. Does the fire chief support community education? Is the fire chief providing money, time, and people for the program? Do the fire chief and other senior officers make community education part of day-to-day operations? When others see that community education is part of the department's mission, they will lend their support. Community involvement is needed to be successful.

What is my personal commitment?

Several specific actions are needed to build and maintain an effective planning process and a successful community education program. These actions become a personal action plan.

Do the right things <u>and</u> do things right. It is important to have an understanding of the steps that must be completed in order to establish a sound program. This includes completing a planning process, gaining support of the department leaders, and developing a partnership with the community. Take the time to do quality work. Short cuts seldom work and often take more time to fix than doing it right from the start. Be dependable and willing to do your share.

Invest the required time and effort. Organizing successful community education initiatives takes time, resources, and support. It takes time to build a community partnership and to develop a relationship with the target audience.

Follow a proven successful process to identify and reduce community risk. This guide provides a proven method for conducting a planning process. However, take the time to read journal articles and textbooks on community education. If possible, attend the National Fire Academy (NFA) community education courses, or similar courses taught at the State or local level. It helps to learn several proven methods of planning community education and risk reduction. Get insight into programs that have been less than successful and avoid falling into the same traps.

Collect data and be objective. No one knows all the answers to solving fire and injury problems. Be objective in your decisionmaking and in the use of educational methods. Before guessing about the community's problems and risks collect accurate data. Don't rely on intuition to determine solutions. Instead, meet with the target audience and the other members of the community team. Get as much information as you can from them and then analyze what the information means. If needed, get help in the analysis but don't get stuck in "paralysis from analysis." Step forward, make decisions, and move ahead.

It is easy to become protective of department initiatives and programs, even to the point of excluding other agencies. Be willing to give up sole program ownership and encourage others to share ownership

Be a community education activist. In any great change effort there are always leaders who step forward and lead the charge. That person, the public educator, becomes a cheerleader, an organizer, a promoter, a recruiter, and a coordinator all in one. Get others excited about the possibilities of reducing fires and injuries through community education, and then channel that excitement into action and involvement. Enthusiasm will get others on the team. Continually discuss community education programs with department leaders so they stay committed.

> The toughest task is winning the support of other department members. Take the time to listen and consider their program recommendations and recruit assistance in marketing the program to the community.

Why is organizational support so important?

Organizational commitment is essential for success. There may be some short-term victories without organizational commitment, but long-term, community-changing results will not be possible.

There are several tangible benefits that come with organizational support. First, department resources are available more readily. This includes the most valuable department resource—its people.

Secondly, fire department leaders will help make connections with other leaders in the community. Fire chiefs are active in the community and interact regularly with other leaders. The chief will facilitate introductions to the other leaders so the public educator can promote the community education initiative directly to decisionmakers. In short, the fire chief can open doors easily that otherwise might require a great deal of effort.

Finally, when the organization is supporting community education initiatives publicly it sends a signal to other community organizations and agencies: Reducing fires and injuries in our community is important! When your department supports community education, other community partners will get on board.

What department actions support community education?

Organizational commitment for community education is not a one-time thing. It's an all-the-time effort that becomes a part of the department culture. This level of commitment takes strong leadership from the public safety educator. It is not easy to achieve. These points are key: 1) Community education can reduce fires and injuries effectively when used as part of the department's overall prevention strategy; 2) It may take months or years for community education initiatives to be fully effective; 3) The community looks to the fire department for leadership; and 4) Effective community education requires a partnership.

Make sure the chief understands what community education is, and what is required to have a successful education program. Educate the chief and other decisionmakers and leaders about community education.

The following paragraphs discuss several organizational actions that are important in order for a community education program to be effective and to achieve its goals.

Institutionalize community education as an organizational value. Fire suppression is a program deeply ingrained as a department value. Firefighters don't argue that fire suppression is a key mission of the department. The same should be true for community education. Community fire and life safety education should be included in the mission statement. It should be part of the department's budget. Job descriptions should include community education as a required duty, and personnel should be trained as community educators. Community education should be part of the department's overall prevention strategy.

Commit department resources to the community education program. Community education, just like any other department program, requires resources. Those resources include money and people. Some community education initiatives require a multiyear commitment from the department. For example, school-based education programs such as the NFPA's *Risk Watch* must be used in the classroom over a period of years to be effective.

Identify the short- and long-term resources required for any community education program. Make decisionmakers aware of all the resources required for the program when gaining their support. This information allows decisionmakers to budget to meet ongoing needs.

The public demands that fire suppression crews be staffed adequately and have the apparatus and equipment necessary to put out fires and respond to EMS emergencies. The community educator should avoid demanding the same funding as suppression: This will alienate the fire chief and operational personnel. Rather, work with the fire chief to obtain the resources necessary for success while respecting the needs and roles of other department programs, including suppression.

Community education is made a department activity. The fire departments with the most successful community education programs all share one common strategy: every person on the department is a community educator. This includes firefighters, code inspectors, investigators, and even administrative personnel. Everyone must be involved in the overall community education program. When this happens a multitude of resources become available.

Decisionmakers and senior officers are taught the process of fire and injury prevention. Senior officers should receive training on how to develop a prevention strategy which includes conducting a community risk analysis, identifying the appropriate strategies to reduce the problems, and methods for involving the community in the prevention program.

Senior officers and department leaders find it difficult to support what they don't understand. Identify training opportunities such as the NFA's *Community Education Leadership* and *Strategic Analysis of Community Risk Reduction* courses. Similar State and local training programs also may be available.

> Encourage the senior officers and decisionmakers to take these classes as part of their professional development. The lessons and skills they learn will be an asset to you and the community education program.

Prevention strategies include adopting and enforcing up-to-date fire safety codes and standards, and building plans review. A comprehensive prevention program includes community education, adopting fire and life safety codes, code enforcement, and building plans review. Education alone will not be fully effective.

Expand prevention strategies to include injury prevention. In many communities the greatest risk to the health of the public is not fire, but rather more generally preventable injuries. Most departments provide some level of emergency medical services (EMS), so injury prevention should be part of the department's prevention mission.

The community educator should have a good relationship with the fire chief. The educator must make a commitment to building and maintaining a good working relationship with the chief. What does a good relationship involve? Here are some important considerations:

- **Trust.** Part of the relationship with the fire chief is mutual trust. A fire chief must be able to trust the information being provided by an educator. Always be completely truthful, even if it's painful to do so.

- **Mutual Respect.** Respect the fact that there are more programs to consider than just community education. There are times that resources requested are not going to be available because of other department priorities.

- **Acting proactively.** Look to the future and have at least a 2- to 3-year plan for the program. Maintain a community planning process. Advanced planning helps to align department resources more effectively for program support.

- **Seek to understand before being understood.** Take the time to learn about the rest of the department's programs and needs and then relate the education program's needs in those terms.

- **Play by the rules.** It is tempting to bend the rules and to go outside regular channels to get things done. But this strategy breaks down mutual trust. There are always rules and policies. Learn them and use them effectively.

- **Do a few things well.** Identify the highest priorities for the community through a planning process, and then attack those priorities with all the resources available so that the program will be successful.

Summary

Success comes from building internal support for community education, understanding the role the community educator plays in successful programs, and creating an organization that is focused on prevention and community education. The community educator must be the leader in this process, and be responsible for guiding and educating other people and groups. The "five-step" process outlined in this guide is a key to success. It really works!

The public educator must complete the planning process by determining the most serious fire and/or injury problems facing the community and the best solutions. Then he or she must be willing to reach out to the community and work with other agencies as part of a community team. Finally, program results must be evaluated correctly.

This guide provides you with information to develop the skills for conducting a basic but effective planning process. Through the years, the lessons and processes have proved effective. It's now up to you to use the planning process as the foundation for your public education

Step 1: Conduct a Community Risk Analysis

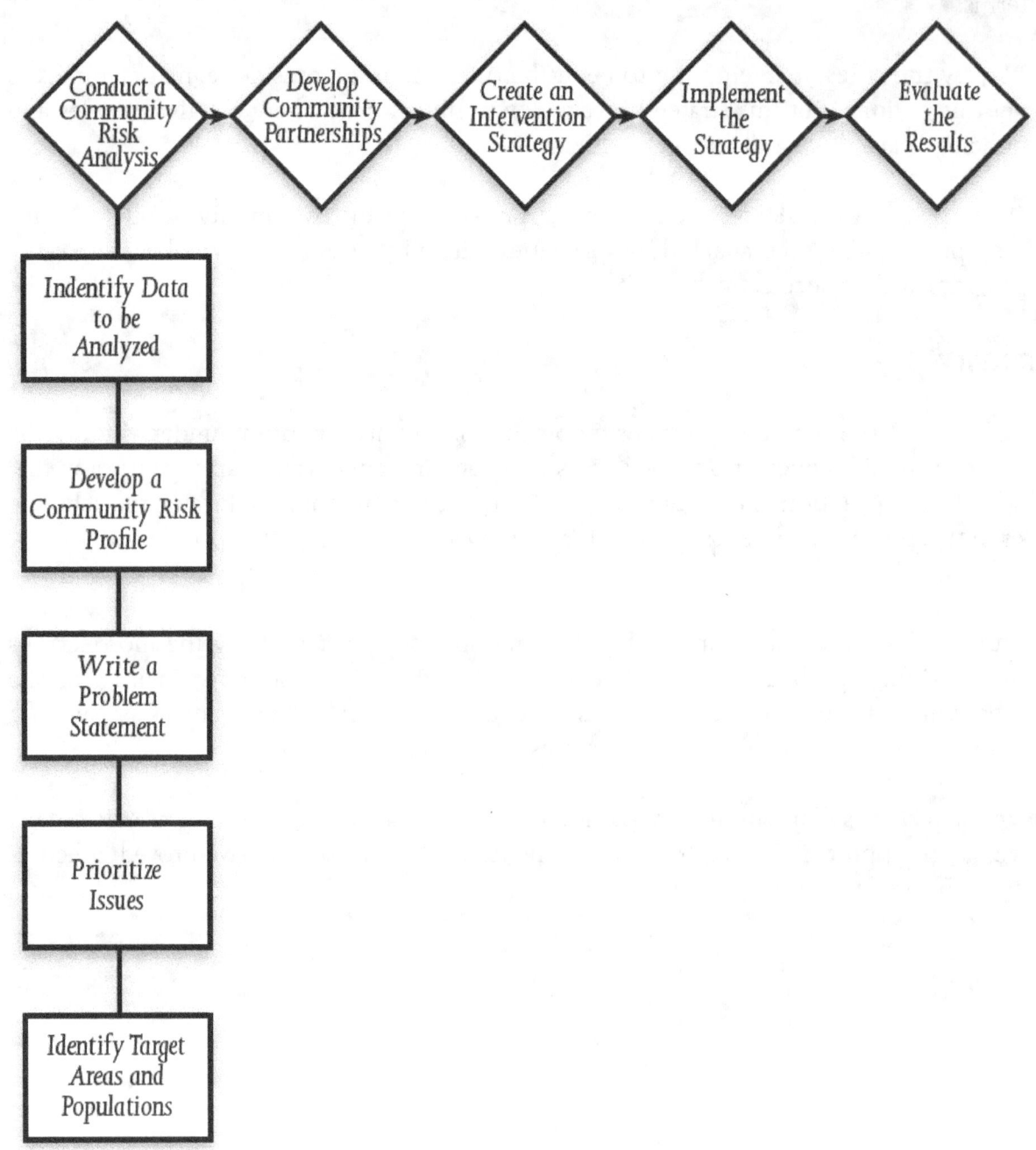

Chapter 1
Step 1: Conduct a Community Risk Analysis

Introduction

What is a community risk analysis? A community risk analysis is a process that identifies fire and life safety problems and the demographic characteristics of those at risk in a community. A thorough risk analysis provides insight into the worst fire and life safety problems and the people who are affected. The analysis results create the foundation for developing risk reduction and community education programs.

Conducting a community risk analysis is the first step toward deciding which fire or injury problem needs to be addressed. Risk analysis is a planned process that must be ongoing, as communities and people are constantly changing.

Why conduct a community risk analysis? Fire and rescue departments must use facts in order to correctly identify leading safety risks that need attention. A community risk analysis will provide these data.

Consider this: A group of people decide to plan a trip. How will they know the best route to take without knowing where their starting point is? Likewise, it is important to know where a community is before beginning a community education program that will reach a risk reduction goal.

Too often, an objective and systematic community risk analysis is a step that is overlooked in the community education process. Many emergency service organizations address risks based on a perceived need for service that isn't really there. An educator may want to teach people about something that is not a problem, perhaps just because a particular subject is interesting. First, it will be hard to get people to participate in an unnecessary program. Second, nothing actually will be accomplished. This approach can be costly in terms of misdirected resources, continued property loss, injuries, and deaths.

Examples

- An organization reads about a high occurrence of electrical fires in a neighboring community and decides to begin a campaign to address these types of fires.

  However, a data analysis of its own community's fire occurrences reveals that it has very few electrical fires. What it does have is a large number of unattended cooking fires.

  Result: Because its campaign is misdirected to electrical fires, cooking fires and related injuries continue.

- An organization located in the northeastern United States reads of swimming pool drownings occurring in the southwestern part of the country. The organization begins a safe swimming campaign.

  If this organization looked carefully at community data, it would identify that only two swimming pool drownings had occurred in the prior 10 years. Further study would have revealed a large number of child pedestrian injuries and deaths.

  Result: Because its campaign is misdirected toward deaths by drowning, child pedestrian injuries and deaths continue.

> It is easy to become distracted from local issues by big media splashes about national happenings. Don't let that happen. Stay focused! Invest the time to conduct an objective community risk analysis. Having facts about the leading causes of risk and the people being affected will set the stage for a successful risk reduction process.

Objectivity means using reliable data to make conclusions based on facts, not beliefs.

What does a community risk analysis include? A community risk analysis includes five important activities:

1. Identify data to be analyzed.

2. Develop a Community Risk Profile.

3. Write a problem statement.

4. Prioritize issues.

5. Identify target areas and populations.

The analysis provides a factual overview of the risk issues, using the information obtained, and should be used to develop a risk reduction proposal. This will produce a picture of what is wrong and who is at risk.

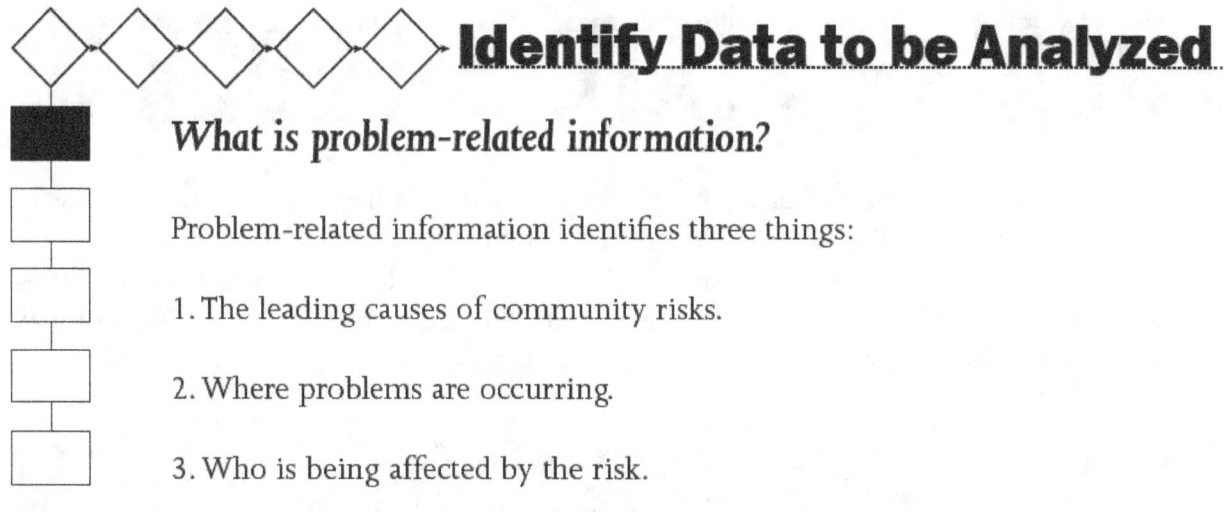

Identify Data to be Analyzed

What is problem-related information?

Problem-related information identifies three things:

1. The leading causes of community risks.

2. Where problems are occurring.

3. Who is being affected by the risk.

Problem-related information is obtained by conducting research, asking questions, and making comparisons based on quality data.

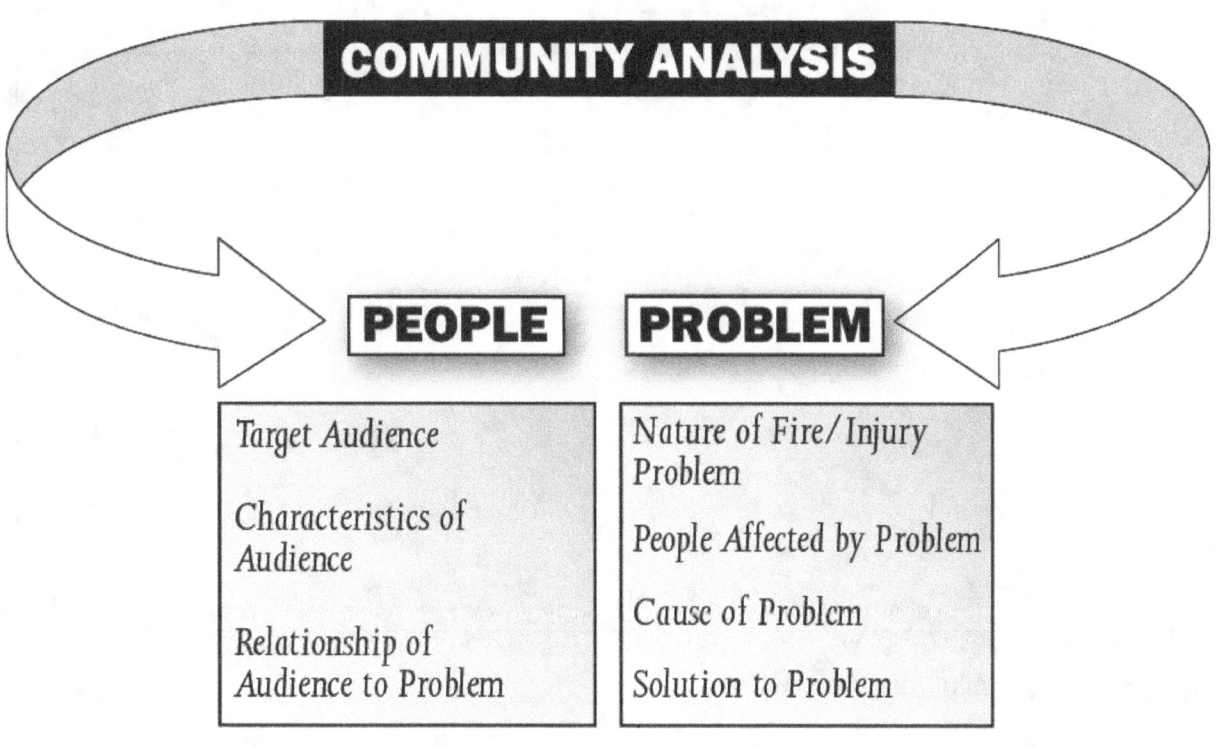

Figure 1: Community Analysis.

How do you locate problem-related information?

Here is an example. A fire department in Safe City, USA, wants to examine the leading cause(s) of fire in the community

LOCAL↔STATE↔FEDERAL

The fire department locates valid data sources to identify the leading causes of fire in America. Sources include the NFPA and the USFA. National injury data also can be sought from the Centers for Disease Control and Prevention (CDC) and the National Safe Kids Coalition. National sources are sought so a comparison can be made to State and local data. The National Fire Data Center Web site provides community educators with published statistical information on the national fire problem.

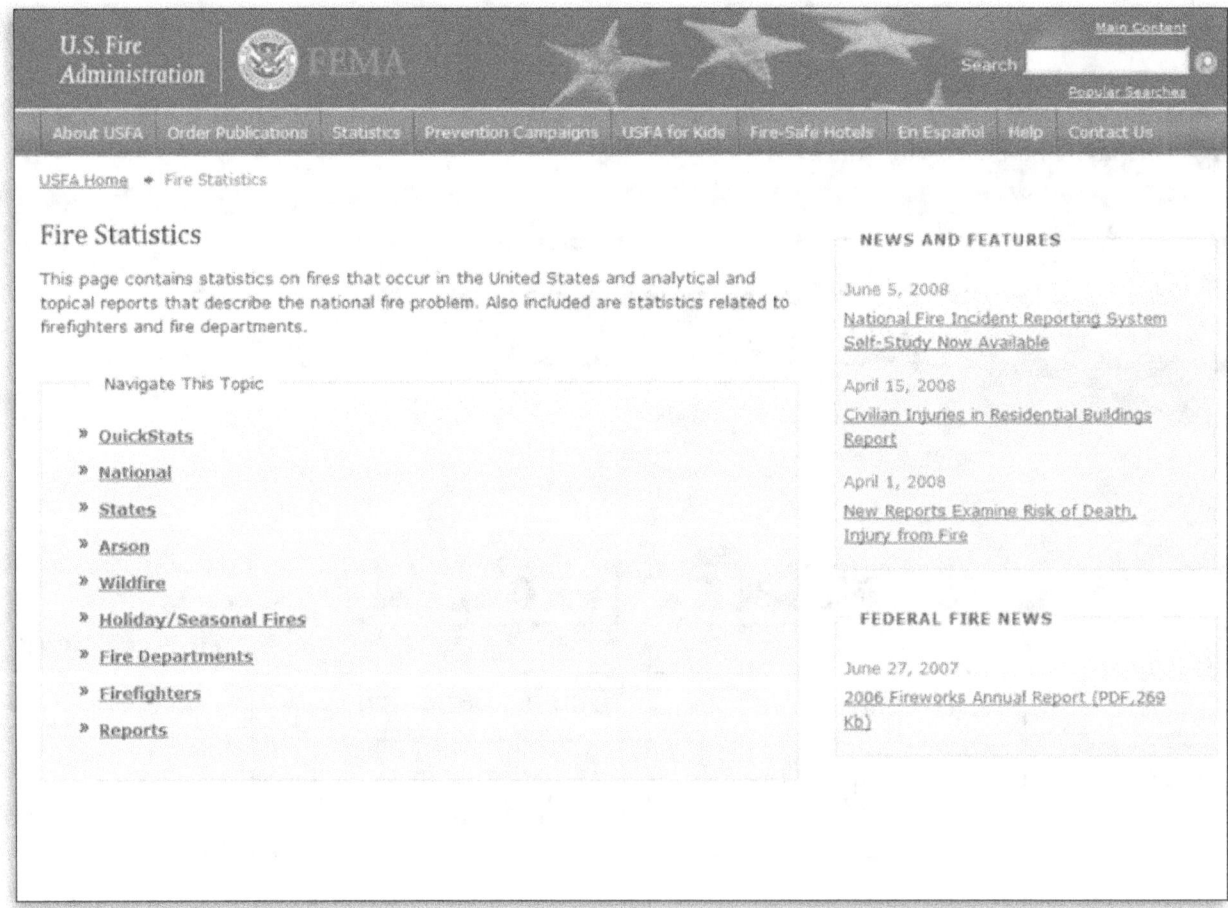

Figure 2: U.S. Fire Administration Web site.

A community educator can log onto the Internet, go to the USFA and NFPA Web sites, and instantly review a profile of the national fire problem.

These organizations are reviewing data regularly from NFIRS. This allows them to build a national profile of how many fires and other incidents occur, note their causes, and indicate where they occur and who is affected. A copy of one page of the NFIRS report form is on the next page.

A
FDID ☆ | State ☆ | MM DD YYYY Incident Date ☆ | Station | Incident Number ☆ | Exposure ☆ | ☐ Delete ☐ Change ☐ No Activity | NFIRS-1 Basic

B Location Type ☆
☐ Street address
☐ Intersection
☐ In front of
☐ Rear of
☐ Adjacent to
☐ Directions
☐ US National Grid

☐ Check this box to indicate that the address for this incident is provided on the Wildland Fire Module in Section B, "Alternative Location Specification." Use only for wildland fires.

Census Tract

Number/Milepost | Prefix | Street or Highway | Street Type | Suffix

Apt./Suite/Room | City | State | ZIP Code

Cross Street, Directions or National Grid, as applicable

C Incident Type ☆
Incident Type

D Aid Given or Received ☆ ☐ None
1 ☐ Mutual aid received
2 ☐ Auto. aid received
3 ☐ Mutual aid given
4 ☐ Auto. aid given
5 ☐ Other aid given
Their FDID | Their State
Their Incident Number

E₁ Dates and Times   Midnight is 0000
Check boxes if dates are the same as Alarm Date
Month Day Year Hour Min
Alarm ☆ — ALARM always required
☐ Arrival ☆ — ARRIVAL required, unless canceled or did not arrive
☐ Controlled — CONTROLLED optional, except for wildland fires
☐ Last Unit Cleared — LAST UNIT CLEARED, required except for wildland fires

E₂ Shifts and Alarms   Local Option
Shift or Platoon | Alarms | District

E₃ Special Studies   Local Option
Special Study ID# | Special Study Value

F Actions Taken ☆
Primary Action Taken (1)
Additional Action Taken (2)
Additional Action Taken (3)

G₁ Resources ☆
☐ Check this box and skip this block if an Apparatus or Personnel Module is used.
	Apparatus	Personnel
Suppression		
EMS		
Other		
☐ Check box if resource counts include aid received resources.

G₂ Estimated Dollar Losses and Values
LOSSES: Required for all fires if known. Optional for non-fires. | None
Property $ ☐
Contents $ ☐
PRE-INCIDENT VALUE: Optional.
Property $ ☐
Contents $ ☐

Completed Modules
☐ Fire-2
☐ Structure Fire-3
☐ Civilian Fire Cas.-4
☐ Fire Service Cas.-5
☐ EMS-6
☐ HazMat-7
☐ Wildland Fire-8
☐ Apparatus-9
☐ Personnel-10
☐ Arson-11

H₁ ☆ Casualties ☐ None
	Deaths	Injuries
Fire Service		
Civilian		

H₂ Detector
Required for confined fires.
1 ☐ Detector alerted occupants
2 ☐ Detector did not alert them
U ☐ Unknown

H₃ Hazardous Materials Release ☐ None
1 ☐ Natural gas: slow leak, no evacuation or HazMat actions
2 ☐ Propane gas: <21-lb tank (as in home BBQ grill)
3 ☐ Gasoline: vehicle fuel tank or portable container
4 ☐ Kerosene: fuel burning equipment or portable storage
5 ☐ Diesel fuel/fuel oil: vehicle fuel tank or portable storage
6 ☐ Household solvents: home/office spill, cleanup only
7 ☐ Motor oil: from engine or portable container
8 ☐ Paint: from paint cans totaling <55 gallons
0 ☐ Other: special HazMat actions required or spill > 55 gal (Please complete the HazMat form.)

I Mixed Use Property ☐ Not mixed
10 ☐ Assembly use
20 ☐ Education use
33 ☐ Medical use
40 ☐ Residential use
51 ☐ Row of stores
53 ☐ Enclosed mall
58 ☐ Business & residential
59 ☐ Office use
60 ☐ Industrial use
63 ☐ Military use
65 ☐ Farm use
00 ☐ Other mixed use

J Property Use ☆ ☐ None
Structures
131 ☐ Church, place of worship
161 ☐ Restaurant or cafeteria
162 ☐ Bar/tavern or nightclub
213 ☐ Elementary school, kindergarten
215 ☐ High school, junior high
241 ☐ College, adult education
311 ☐ Nursing home
331 ☐ Hospital

Outside
124 ☐ Playground or park
655 ☐ Crops or orchard
669 ☐ Forest (timberland)
807 ☐ Outdoor storage area
919 ☐ Dump or sanitary landfill
931 ☐ Open land or field

341 ☐ Clinic, clinic-type infirmary
342 ☐ Doctor/dentist office
361 ☐ Prison or jail, not juvenile
419 ☐ 1- or 2-family dwelling
429 ☐ Multifamily dwelling
439 ☐ Rooming/boarding house
449 ☐ Commercial hotel or motel
459 ☐ Residential, board and care
464 ☐ Dormitory/barracks
519 ☐ Food and beverage sales

936 ☐ Vacant lot
938 ☐ Graded/cared for plot of land
946 ☐ Lake, river, stream
951 ☐ Railroad right-of-way
960 ☐ Other street
961 ☐ Highway/divided highway
962 ☐ Residential street/driveway

539 ☐ Household goods, sales, repairs
571 ☐ Gas or service station
579 ☐ Motor vehicle/boat sales/repairs
599 ☐ Business office
615 ☐ Electric-generating plant
629 ☐ Laboratory/science laboratory
700 ☐ Manufacturing plant
819 ☐ Livestock/poultry storage (barn)
882 ☐ Non-residential parking garage
891 ☐ Warehouse

981 ☐ Construction site
984 ☐ Industrial plant yard

Look up and enter a Property Use code and description only if you have NOT checked a Property Use box.
→ Property Use | Code
Property Use Description

NFIRS-1 Revision 01/01/05

Figure 3: NFIRS Form.

The fire department locates data sources to identify the leading causes of fire in the State. State data, usually available from the State fire marshal's office, should be used to make a comparison to national and local data.

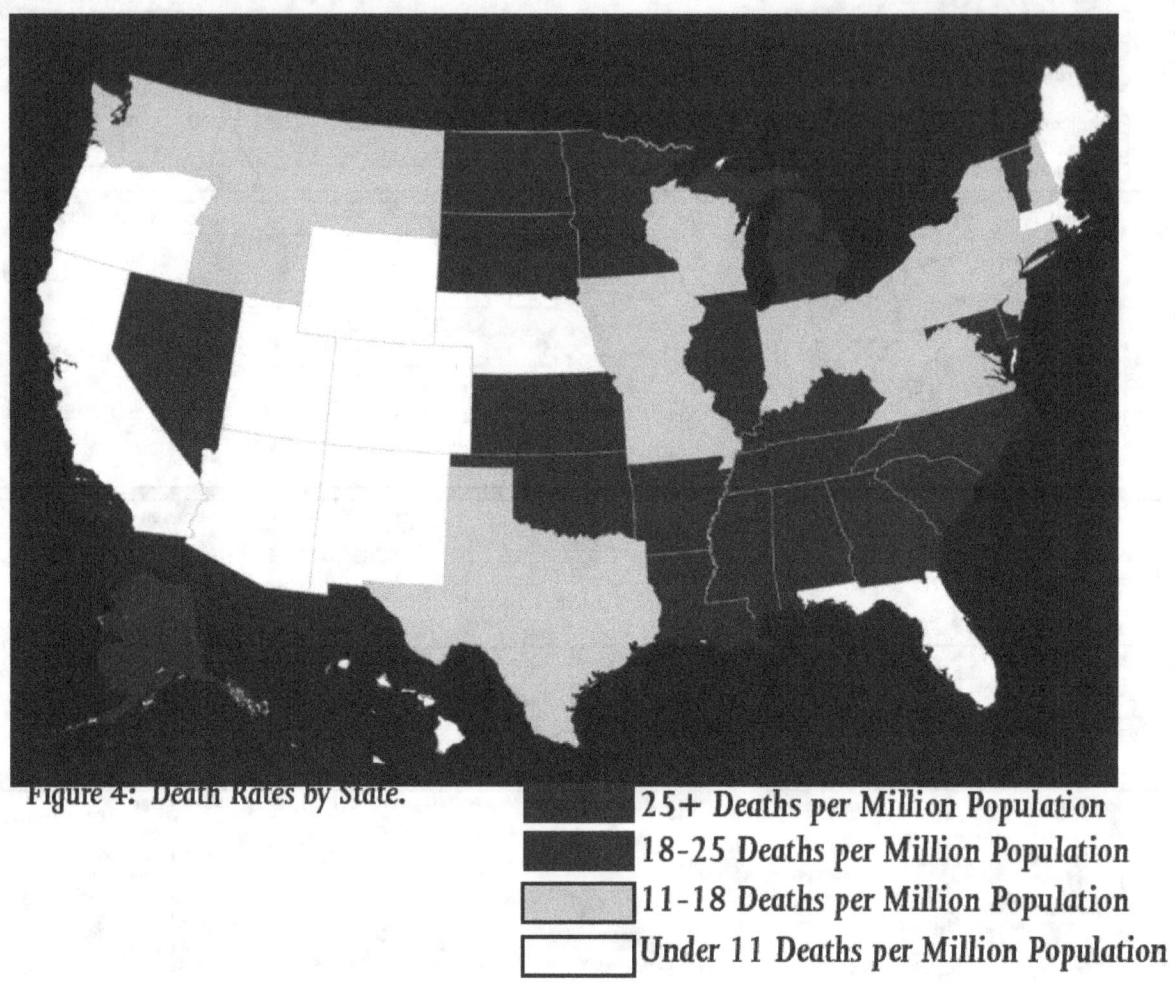

Figure 4: Death Rates by State.

◼ 25+ Deaths per Million Population
◼ 18-25 Deaths per Million Population
▨ 11-18 Deaths per Million Population
☐ Under 11 Deaths per Million Population

Note: 1995 data from Indiana and Nevada are incomplete. 1994 data are shown.
Sources: State Fire Marshals and the United States Fire Administration.

Some examples of State data include a comparison of your State's fire deaths to those of the whole United States. This information will provide an indicator of the magnitude of the problem in your State. The bar chart in Figure 5 illustrates such a comparison.

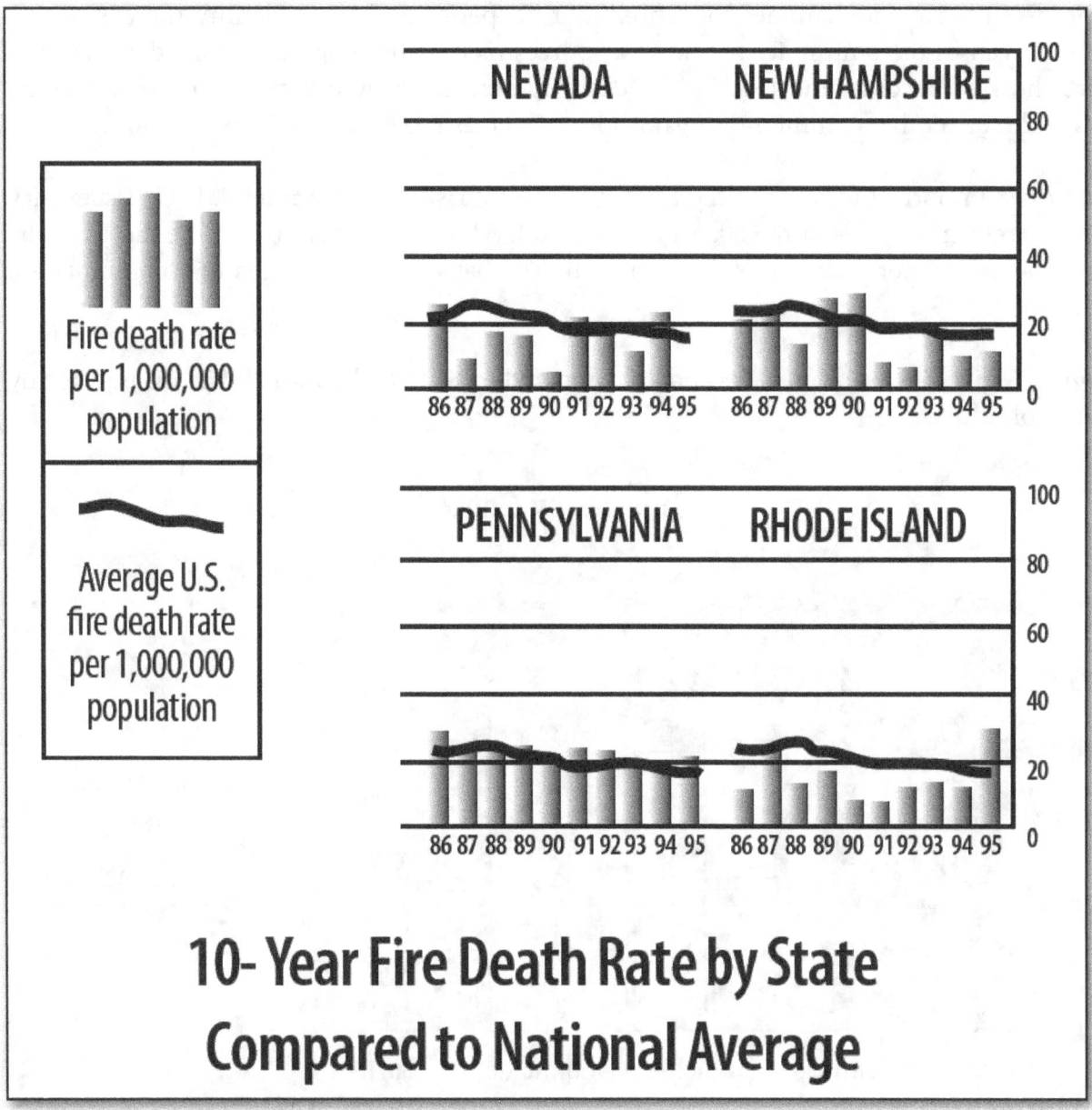

Fire death rate per 1,000,000 population

Average U.S. fire death rate per 1,000,000 population

10- Year Fire Death Rate by State Compared to National Average

Figure 5: Comparison of State and National Fire Deaths by Year.

The fire department examines local fire/injury experience data to identify the extent and leading causes of community fire/injury. Sources include fire department incident records, local hospital records, and local health department records. Additional injury data may be available from other community organizations, such as the local Safe Kids Coalition.

It is most important to identify the leading causes of risk at the local level. Local issues may be different from those at the State and national level. Local data are by far the most important source of information. It is through local data that specific risks in your community are identified.

Figure 6 gives an example of local data compared to State and national data on injuries by cause of injury.

Injuries by Cause

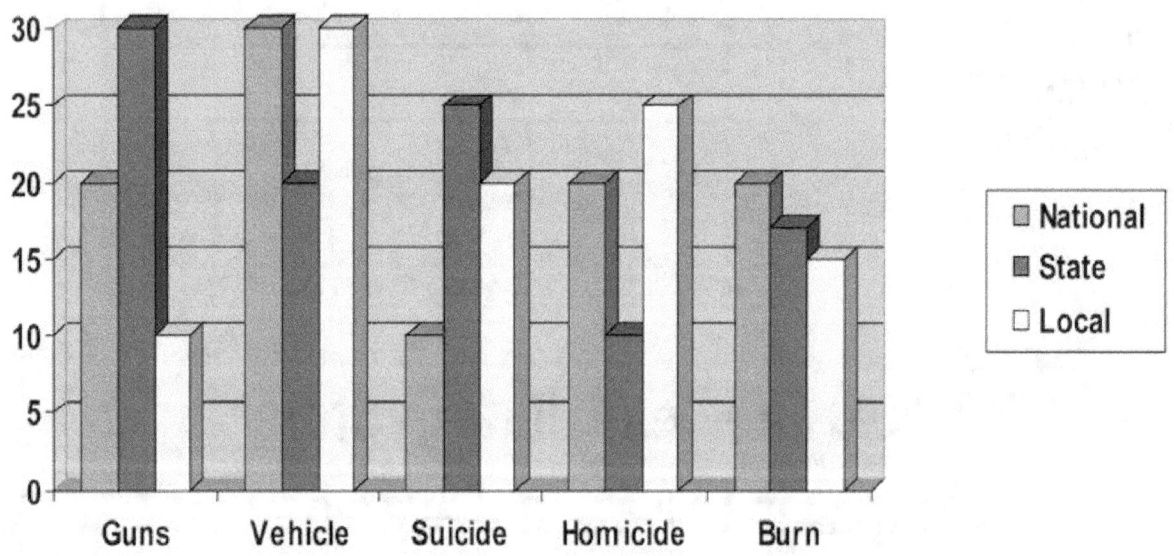

Figure 6: Local Injuries by Cause.

The goal is to identify the leading causes of fire/injury at the national, State, and local level. It is important to examine all levels so that comparisons can be provided to potential partners from the local community.

If an organization does not have a formal data collection system, it is never too late to start one! Potential sources to assist with learning how to collect data are listed at the end of this section.

How much local problem information is needed?

Quite a bit of information is required in order to draw accurate conclusions. Even though national and State information is available, it is critical to develop the same type of profile at the local level.

Examine at least 3 years (more if possible) of data. This will allow the identification of a baseline (average) of how many times a specific incident occurs annually. The smaller the community, the further the history search should be extended.

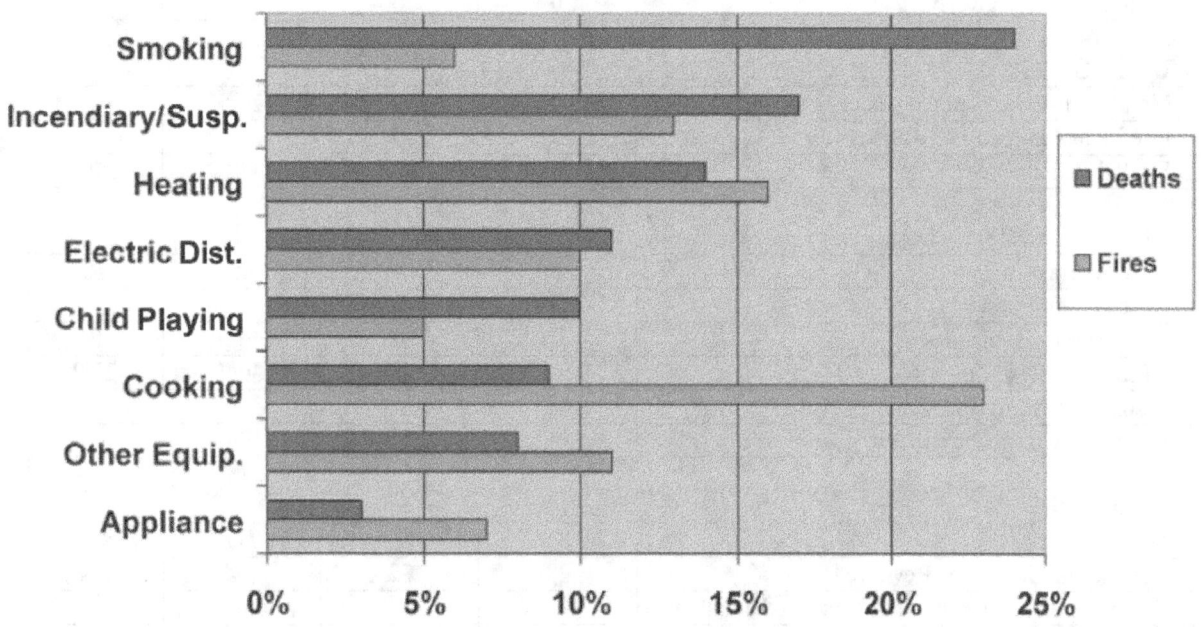

Leading Causes of Home Fires and Deaths
1993-1997 Annual Averages

Figure 7: Annual Average Comparison of Home Fires and Deaths.

It is easy to create a simple database from existing run sheets. This allows the fire department to identify fire and injury causes and their occurrences.

For example the following activity report can be used to study the fire problem in your city.

Activity Report for August 2001
Fire Department

	Aug. 2001	Aug. 2000	Year to Date 2001	Year to Date 2000
SUPPRESSION				
Total Incidents	150	156	1,218	1,102
Estimated Loss	6,102	265,905	854,642	657,654
Multiple Alarm Fires	0	1	1	2
Average Number of Firefighters Per Incident	7	7	8	7
Average Response Time	2:41	2:24	2:38	2:44
Average Time Spent Per Incident	19:16	23:38	22:36	26:56
Civilians Injured	0	1	7	5
Civilian Fatalities	0	0	0	0
Civilians Rescued/Revived	0	0	0	1
TRAINING				
Monthly In-House Training Hours Provided	0	0	20.5	26.5
Firefighters Injured	0	0	6	2
Firefighter Fatalities	0	0	0	0
INVESTIGATIONS				
False Alarms	6	4	48	40
False Alarm Arrests	0	0	1	2
False Alarm Convictions	0	0	1	2
Incendiary Fires/Fireworks	5/0	6/0	30/0	23/0
Arson Arrests/Fireworks	1/0	6/0	21/0	17/0
Arson Convictions/Fireworks	1/0	2/0	19/0	10/0
INSPECTIONS				
Total Inspections	84	125	711	848
Site and Building Plans Reviewed	28	22	164	174
Citizen Complaints Received about Fire Violations	1	0	6	6
Number of Appeals	0	0	0	1
Legal Action Taken	0	0	0	0
PREVENTION/EDUCATION				
Free Smoke Detectors Installed	15	6	215	236
Public Education Programs (Children's Village - 1; School - 0; Other 1)	9	11	134	131
People in Attendance (Children's Village - 30; School - 0; Other - 30)	450	340	5,890	4,935
Media Articles/Programs	6	15	74	90

Figure 8: Local Activity Report.

Safe City Example

The Safe City, USA, fire department responds to 100 structure fires each year. The fire department decides to examine 10 years of structure fire experience data to identify the leading structure fire cause in Safe City.

The Safe City Fire Department discovers it has responded most often to fires resulting from unattended cooking: 400 such calls over a 10-year period. To determine the average number of cooking fires per year, divide the total number of cooking fires in the 10-year study (400) by the number of years in the study (10). The result: Safe City Fire Department experiences an average of 40 cooking fires per year.

These data can be restated as 40 cooking fires per 100 yearly structure fires (40/100, or 40 percent.

Safe City Fire Cause

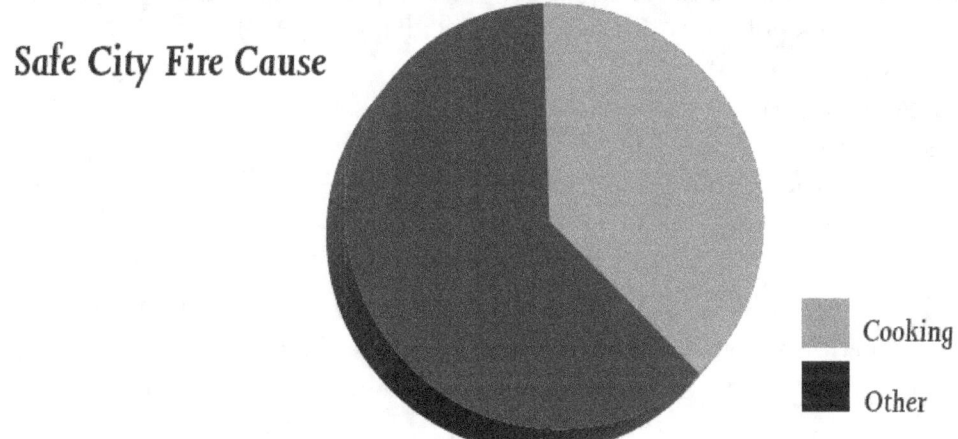

Cooking

Other

Although 40 cooking fires each year may not seem like a large number, comparing the occurrence with all other fire causes identifies it to be the leading structure fire problem. Cooking fires also are a frequent cause of burn injuries.

By completing this data analysis, Safe City Fire Department has identified the following information:

- The leading cause of structure fire is unattended cooking.

- The fire department is most likely to decrease the number of structure fires in its community by developing a campaign to reduce cooking fires.

- The baseline average number of cooking fires in Safe City is 40 per year. By comparing next year's incidence of cooking fires to this baseline number, the fire department will be able to evaluate the progress of its campaign to reduce unattended cooking.

> **Examining at least 3 years of data (more if possible) will provide a more accurate profile of leading risk issues.**
>
> **Examining less than 3 years of data will not provide an accurate profile of leading risk issues, and the real issues may be overlooked.**

What type of local problem-related information is needed?

To conduct an objective risk analysis, the following types of local information must be obtained:

- types of incidents (fires, falls, etc.);
- root factors leading to incident (how does a typical incident happen?);
- how often the incidents occur;
- location of the incidents (geographic distribution);
- when incidents occur (time, day, month);
- cost of the incidents (expense, injuries, loss of life); and
- types of incidents happening most frequently.

This type of local information can be displayed like the examples on the following page in order to analyze the information and make decisions about local problems.

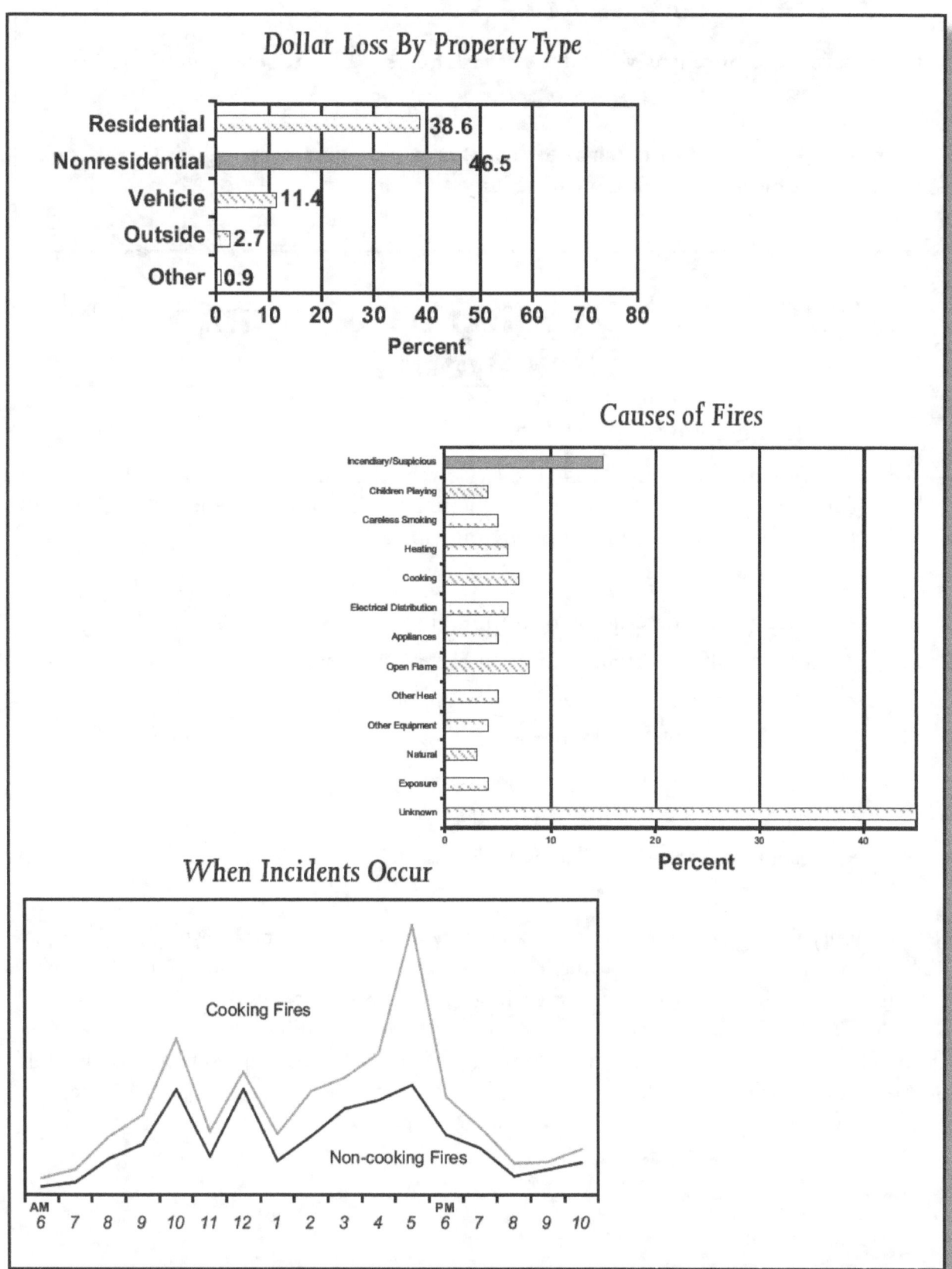

Figure 9: Examples of Local Data Displays.

> Conducting a comprehensive analysis of risk helps to ensure selection of the "priority" issues.
>
> Examining only one area (such as expense) may not provide enough information to make an objective decision on what issue to address.

Develop a Community Risk Profile

What is a community risk profile?

A community risk profile is an overview of the information gathered about the community. It can serve as an introduction to a problem statement. A community profile should include the following information:

- demographic description of community;
- demographic distribution of high-risk populations;
- brief description of selected fire/injury risk issue;
- public perception of the problem;
- existing myths about the problem;
- political support for the issue; and
- resources currently available to address the risk issue;

A community profile does not have to be a lengthy report. A paragraph or two for each section is enough. The community profile leads a reader to the problem statement. Several profiles may need to be developed, one for each targeted risk selected.

Data collected for the local area can be summarized in charts that represent a demographic profile. The following chart details the demographics for one county in terms of age, sex, and income.

Profile of Health Status Indicators

Demographics

Males | | | | | | ### Females | | | | |

COUNTY	# ALL RACES	% WHITE	% AFRICAN-AMERICAN	% NATIVE AMERICAN	% ASIAN	COUNTY	# ALL RACES	% WHITE	% AFRICAN-AMERICAN	% NATIVE AMERICAN	% ASIAN
ALL AGES	2,961	98.5	0.1	1.3	0.0	ALL AGES	2,964	98.2	0.1	1.6	0.1
< 1	41	93.9	0.7	5.4	0.0	< 1	33	98.3	1.0	0.0	0.7
1 to 4	147	93.5	0.5	6.0	0.0	1 to 4	154	93.2	1.1	5.2	0.5
5 to 9	246	98.0	0.0	2.0	0.0	5 to 9	203	97.0	0.0	3.0	0.0
10 to 14	280	99.3	0.0	0.7	0.0	10 to 14	268	98.9	0.0	1.1	0.0
15 to 19	217	97.7	0.0	2.3	0.0	15 to 19	198	98.0	0.0	2.0	0.0
20 to 44	935	98.2	0.2	1.5	0.1	20 to 44	886	97.5	0.2	2.2	0.1
45 to 64	657	99.8	0.0	0.2	0.0	45 to 64	648	98.9	0.0	1.1	0.0
65 to 74	243	99.6	0.0	0.4	0.0	65 to 64	294	99.7	0.0	0.3	0.0
75 +	195	100.0	0.0	0.0	0.0	75 +	279	100.0	0.0	0.0	0.0

STATE	# ALL RACES	% WHITE	% AFRICAN-AMERICAN	% NATIVE AMERICAN	% ASIAN	STATE	# ALL RACES	% WHITE	% AFRICAN-AMERICAN	% NATIVE AMERICAN	% ASIAN
ALL AGES	1,586,887	83.0	7.7	8.1	1.2	ALL AGES	1,666,742	82.9	7.7	8.1	1.3
< 1	21,371	78.4	10.1	10.0	1.5	< 1	20,090	76.8	10.4	10.4	2.4
1 to 4	100,081	78.0	10.3	10.0	1.7	1 to 4	96,344	77.5	10.6	10.3	1.6
5 to 9	122,245	77.6	10.1	11.1	1.2	5 to 9	116,596	77.3	10.2	11.1	1.4
10 to 14	132,867	77.9	9.3	11.5	1.3	10 to 14	124,696	77.1	9.7	11.8	1.4
15 to 19	123,129	78.9	9.3	10.5	1.3	15 to 19	116,201	78.6	9.6	10.5	1.3
20 to 44	590,427	82.5	8.2	7.6	1.7	20 to 44	589,875	82.0	8.4	7.9	1.7
45 to 64	317,486	87.8	5.2	6.2	0.8	45 to 64	340,946	86.9	5.6	6.4	1.1
65 to 74	109,361	89.8	4.3	5.5	0.4	65 to 74	133,876	89.0	4.9	5.6	0.5
75 +	70,141	90.0	4.5	5.3	0.2	75 +	128,773	90.0	4.6	5.2	0.2

Income/Poverty

Income Levels	County Number	State Number	County Rate	State Rate
Income < 100% Poverty	523	527,088	8.8%	16.25%
Income < 150% Poverty	1,170	884,987	19.7%	27.2%
Income < 185% Poverty	1,646	1,070,444	27.8%	32.9%
Income < 200% Poverty	1,895	1,242,886	32.0%	38.2%
Poverty by Age Groups				
Age 0-17	162	189,553	9.7%	21.7%
Age 18-64	243	247,597	7.5%	14.2%
Age 65-74	31	32,837	5.9%	13.5%
Age 75 +	100	58,620	21.2%	24.1%
Income				
Per Capita Income	$18,667	$17,610	--	--
Median Family Income	$31,279	$28,554	--	--
Unemployment				
Unemployment Rate	67	57,867	2.4%	3.6%
Household				
# Female-Headed Households	74	87,945	--	--
Nursing Facilities			County Number	State Number
Beds per 1,000 Elderly (Age > 75 yrs)			131	185
Patient Rate per 1,000 Elderly (Age > 75 years)			115	144

Figure 10: Local Demographics Example.

Safe City Example
Community Risk Profile

Safe City, USA, population 20,000, is a small but densely populated community located in the southeastern United States. The town grew from a small village to its present size beginning in the early 1900s. Most homes are more than 30 years old and are of masonry construction.

Although the railroad industry once thrived in Safe City, most residents now commute to larger cities for work.

Safe City is an aging community, in terms of both people and properties. A significant percentage (30 percent) of Safe City's citizens are retired. Many reside in city-owned housing units. A significant portion of the older adult population has difficulty with mobility, hearing, or sight. Most live on a fixed income.

Unattended cooking is the leading cause of fire and fire-related injury in Safe City: 40 percent of both structure fires and related injuries are caused by this risk. Half of the fires (an average of 20 each year) occur in the homes of older adults living in city-owned housing units.

During a recent survey conducted by the fire department, only 10 percent of the older adult population could identify (or guess) that cooking was the leading cause of fires in Safe City. Most of the older adults stated they were safe from fire while in their homes.

The community of Safe City is very supportive of its older adult population. Although progress on issues related to the elderly is sometimes slow, the local housing authority plans, in response to projected community demographics, to double the number of housing units for older adults over the next 15 years.

Write a Problem Statement

What is a problem statement?

A problem statement provides a fact-based overview of the problem and who is affected by it. It also provides a vision of what the organization proposes to do about the problem.

Here is an example of a problem statement developed in Safe City:

Problem Statement

Unattended cooking causes 40 percent of fires and fire-related injury in our community. Half of these cooking fires (an average of 20 each year) occur in the homes of older adults who reside in city-owned housing units. Interviews with fire victims identify that most fires occur because the cooking process has been left unsupervised.

The Safe City Fire Department proposes the development of a community-based campaign to address cooking fires among the older adult population living in city-owned housing units. This proposal is based on the following factors:

- There is a high occurrence of cooking fires and fire-related injuries at the complex.

- The community and housing authority are supportive of the older adult population.

- There is a realistic possibility that such an intervention will be successful.

- A successful campaign can serve as a model for future citywide efforts.

Done well, the community risk profile and problem statement will provide a factual rationale of why the selected risk issue should be addressed.

Share the document with the community and convince others to join the team! A poorly written problem statement (lacking facts) may damage your organization's credibility and result in less support for your proposed effort.

Prioritize Issues

How should the priority risk issue be decided?

Making an objective decision on which risk issue to address is a process that takes time, effort, and patience. It is a decision that must be governed by local need. Following an organized process leads to selecting the risk issue that most needs attention as the priority risk issue.

These factors can influence a fire department's decision to select a particular risk issue:

- high injury/fatality statistics;
- high dollar loss; and
- rapid increase in frequency of occurrence.

Once an idea has developed about a particular risk issue. It is time to examine the second phase of community risk analysis: people and community information.

Relying solely on problem-related data wouldn't provide enough information to formulate a risk reduction campaign. Additional factors should be considered.

What is people- and community-related information?

People- and community-related information is based on demography. Simply stated, demography is the study of people and their environments.

Examples of demographic information:

- housing information: number of homes in the community; types of properties; owner occupied/rental units; general condition of structures; age of homes, etc.;
- people information: number of residents; population density; age/gender/education levels; family structures; disability types, etc.; and
- economic information: family incomes; individual median income; overall community economic projections; number of people in poverty; number of people getting financial assistance, reduced rent, or free or reduced lunch.

Demographic information can be gleaned from both formal and informal sources.

Formal sources may include local planning and community development offices, the Census Bureau, the Chamber of Commerce, economic development offices, hospitals, and health departments.

Informal sources may include local schools and Head Start centers, women's shelters, assisted living facilities, day care centers, community associations, social services, and reports by clergy and established community leaders.

Comparing problem-related information with community demographics helps identify who is most at risk, what factors place them at risk, and where a risk reduction campaign should be directed.

Safe City Example

Safe City, USA, identified these facts about its cooking fire problem:

- Most cooking fires (50 percent, in fact) occur in the homes of older adults with limited incomes who reside in city-owned housing units.

- Most housing units are located in the center portion of the community.

Invest the time and effort to learn about the community. Find out who is most at risk from the selected risk issue. The success of the program will depend upon it.

Omitting this step (or not performing a quality analysis) can result in future efforts not being directed at the correct target population.

Identify Target Areas and Populations

What factors place people at greater fire/injury risk?

Some population types have a disproportionately higher rate of morbidity and mortality. (Morbidity refers to injuries and mortality refers to deaths.) Several factors increase the risk level of a population segment:

- Age: children under age 5 and adults over age 65.

- Disability, including cognitive and developmental, mobility, visual, and hearing.

- Sociocultural and economic status.

- Gender: Males historically exhibit higher fire and injury morbidity and mortality rates.

- Language and communications barriers.

A disability can affect a person's ability to react and respond effectively to an emergency in progress. Disabilities can affect any population.

Cognitive and developmental disabilities refer to a person's inability to process information. Mobility refers to a person's ability to move about. Vision and hearing refer to the ability to process sight and sound. Poverty generally increases the risk of fire and injury.

Social and cultural diversity are realities in all communities. Changing family structures, peer influences, and language diversity are all examples of social and cultural issues that must be considered when planning to address risk.

A person's vulnerability to risk can dramatically increase when several of the above-listed factors apply to that person.

Example: A visually-, mobility-, and hearing-impaired older adult living in poverty will be at substantially increased risk of unintentional injury or fire.

Once a specific risk issue has been identified, the affected populations profiled, and the location of occurrences plotted, the next logical step is to learn why the risk exists.

This is the time to consider the sequence of events and root factors that led to the existence of the selected risk event.

> **Several national organizations have conducted research on high-risk populations. Apply their conclusions to your community. References are listed at the end of this section.**
>
> **Making assumptions about high-risk populations can be a dangerous strategy. Guessing which interventions will work among various groups of people may result in the inability to reach this audience effectively.**

Why examine the events and root causes that led to the problem?

An effective risk reduction intervention requires investing the time to obtain facts about what is causing the local risk in order to create an objective community profile and problem statement.

Event sequencing and root factor exploration can be conducted in two ways:

1. If a good incident reporting system is available, the data may be obtained there.

   Example: NFIRS 5.0 reporting software includes a section for human actions contributing to an event.

2. If reporting system data are not available, an organization may decide to conduct post-incident interviews with victims to determine actions that led to an event.

Interviewing people who have experienced a fire or injury provides valuable insight into why it is occurring.

Effective research is accomplished by asking the question "Why?"

Safe City Example

The Safe City Fire Department found that older adults who left food cooking unattended on the stove caused most cooking fires. The fire department interviewed 25 older adults who had experienced this type of fire.

A profile of the typical sequence of events and root factors leading to the local cooking fire problem identified the following elements:

- A person begins the cooking process, usually on the stovetop, and turns the heat on high. Why? Poor eyesight causes the person to have trouble seeing the control setting.

- The person leaves the kitchen while cooking for various reasons (telephone, television, bathroom, etc.). Why? The person leaves the room intending to come right back but gets sidetracked and forgets the stove is on (e.g., starts watching television or talking on telephone).

- The unattended pan ignites on stovetop. Why? Heat is set at high temperature.

- The smoke alarm alerts the person to the fire. Why? The alarm is in working condition and the person hears the alarm.

- The person returns to kitchen and responds inappropriately by removing the flaming pan from the stove and carrying it away. Why? The person does not know the appropriate response to a stovetop fire, which is to turn off the heat and cover the pan with a lid.

- The person is burned. Why? The person drops the flaming pan at his/her feet and spills the burning food.

- The fire department is notified. Why? The person is hurt and realizes the fire is now out of control.

Fire and injury prevention can be studied using a scientific approach. The process of developing a sequence of events and identifying the root cause shows how and why risk events are understandable, predictable, and preventable.

Summary

A community risk analysis identifies fire and life safety problems and the demographic characteristics of those at risk in a community. A risk analysis provides insight into the most significant fire and life safety problems and the people who are affected by them. The results create the foundation for developing risk reduction and community education programs.

Step 2: Develop Community Partnerships

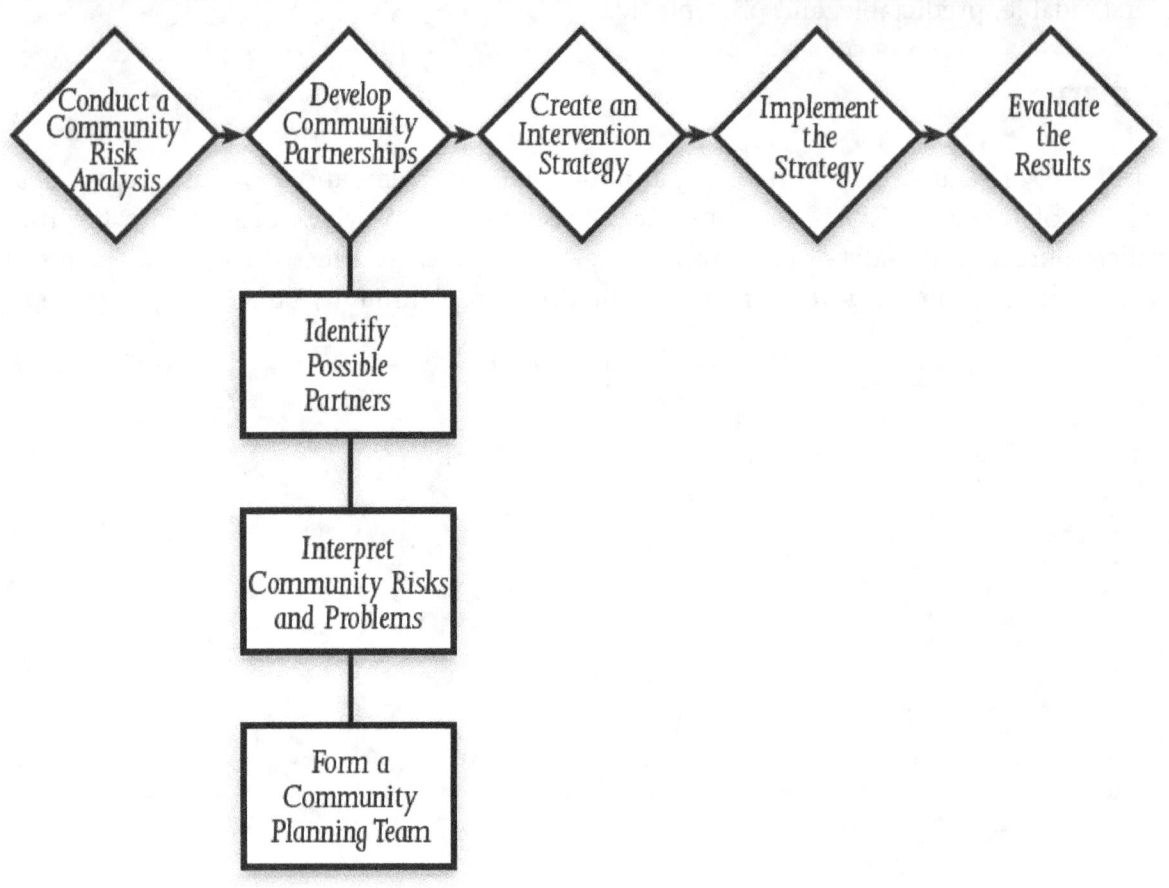

Chapter 2
Step 2: Develop Community Partnerships

Introduction

There are many reasons for developing community partnerships. The main reason should be obvious. Consider this example: There is a major forest fire in a remote area. A single fire department is sent to fight it with one engine unit and one firefighter. No matter how much effort is expended, the fire is just too big to be affected in any significant way without additional help. The same is true when attacking a significant community fire or injury risk. One person or organization will find it impossible to make a positive change without assistance.

Just like fire suppression, public education aimed at reducing community fire and injury risk is serious business. When suppressing fires, most departments rely on mutual aid from other organizations. The same strategy should be applied to community risk reduction programs.

Having one or more partners makes sense in terms of having more minds, bodies, and resources to tackle the problem. Sharing the tasks of targeting the problem, developing a plan, and implementing the solution is more productive in terms of creativity, credibility, and overall effectiveness. The most successful risk reduction efforts are those that involve the community in the planning and solution process.

> **Think teamwork! As with firefighting, and even baseball, community risk reduction is best done by a group working together.**
>
> **The work of a single organization aimed at reducing risk will cost a lot over the long haul in terms of money, time, and effort spent, with little to show for all the effort.**

What community resources are needed for risk reduction?

When the term "resources" is used, many people immediately think of money. It is true that financial funding is needed to reduce most risks. However, money alone will not accomplish anything. A broader spectrum of resources is required to conduct a successful risk reduction effort. Examples of additional important resources include the following:

Knowledge: Knowledge of the problem and possible solutions, including a factual description of the community and its associated risks, is very important. Such knowledge is needed for an accurate community risk analysis.

There are always many people and groups that will be able to offer insight into the community, the people who live and work in the community, and the risks found in the community.

It is up to the community educator to identify them and work with them to reduce the risks.

In-kind support: This involves resources provided in lieu of money, including equipment, printed materials, supplies, and personal effort. Donated professional services such as consultations, program evaluations, and graphic design work are all examples of in-kind resources.

Political support: Support for the risk reduction process from elected officials is very helpful to finding success. The public educator should discuss any plans or ideas with elected officials who have a stake in the program outcome. These people may know how additional resources can be obtained, and often will have information important to program development.

Community support: Partnerships with people and organizations in the community can identify risks, planning measures, and resources required to get the risk reduction process started and the program implemented.

Emphasis is often placed on finances, but additional resources are equally important.

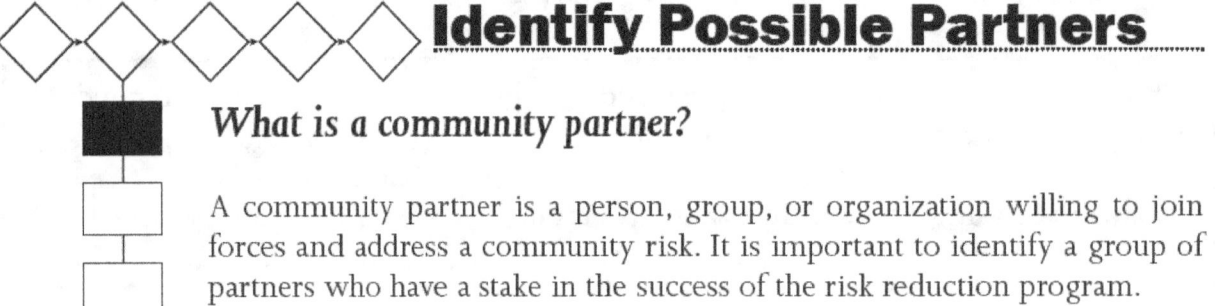

Identify Possible Partners

What is a community partner?

A community partner is a person, group, or organization willing to join forces and address a community risk. It is important to identify a group of partners who have a stake in the success of the risk reduction program.

How does the public educator find help?

An easy way to start finding help is to just start networking and talking to people you know. In addition, the use of the Internet and telephone directory is an easy way to start. A logical first step is to find out which groups might be interested in a specific risk reduction effort or already are addressing the selected risk issue. It is important to avoid duplication of effort.

Think about what part of the population needs to change in order to reduce the risk. Consider which groups are already providing services to these people. Think about those in the community who care about the risk issue and the people it is affecting.

These people may be the best ones to deliver the message. Another approach is to find out who has the resources needed to address the problem.

Bottom line questions: Which people are the movers and shakers in the community who will help get the effort started and the job done? Who can offer leadership, skills, credibility, contacts, influence, and resources?

These are some possible community partners;

- groups already interested or addressing the same or a similar risk issue;
- members of the population who are affected by the risk issue;
- people/groups who feel the financial impact of the risk issue (insurance companies, property owners, American Red Cross disaster services);
- groups already providing services to the population affected by the risk issue;
- community service and advocacy groups; and
- groups that can help deliver messages (media, clergy, schools, marketing organizations).

Recruit a core group of primary stakeholders who will share the responsibilities of developing and implementing a quality risk reduction effort.

Working alone may seem like a good idea; however, failure to obtain the perspective of, and creative input from, others will limit risk reduction efforts.

Interpret Community Risks and Problems

Don't expect everyone in the community to recognize instantly that a particular problem exists. Instead, expect them to be part of the solution process by educating them about the problem. Use the community profile and problem statement as the powerful tool that it is. This work provides a good rationale for why others should consider partnering. Share the vision of the plan and possible solutions with community leaders.

The most effective risk reduction efforts are those that involve the community in the planning and solution process. The community needs to understand and agree that there is a problem and that it can be solved. Sharing the information you have collected in a professional manner will be important. Be sure to explain how and where the information was obtained. Then work with others to continue the process of interpreting the community risk problem.

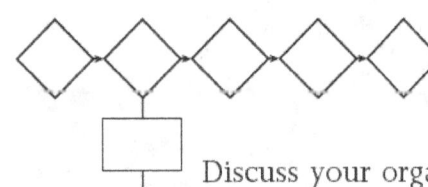

Form a Community Planning Team

Discuss your organization's intention of forming a planning team to reduce the risk problem. Explain what resources your organization can provide.

- Ask the person or group if he/she/it will join the planning team to address the issue.

- Solicit suggestions from people and groups about others who will make good partners.

Allow others the opportunity to provide suggestions.

Trying to state your proposed solution as the only solution will prevent others from offering important insights and suggestions.

Additional benefits of community planning teams

Community planning teams often lead to the development of formal coalitions. A coalition is a group of people with different interests who come together to work on a common problem or issue.

Examples of coalitions that exist in many communities:

- local chapter of the National Safe Kids Coalition;
- local chapter of Mothers Against Drunk Drivers;
- local Crime Stoppers/Solvers Coalition; and
- neighborhood associations.

Coalitions usually include a broad representation of people from the community and lend themselves well to the advancement of political issues. Using integrated prevention interventions to reduce fire and injury risks often requires extensive community and political support.

Summary

Community partnerships are needed in order to reduce any significant fire or injury risk. It is impossible for one person or organization to reduce a serious community fire or injury risk alone. Fire and rescue departments generally rely on mutual aid from other organizations for fire suppression efforts. The same strategy should be applied to community risk reduction programs.

Step 3: Create an Intervention Strategy

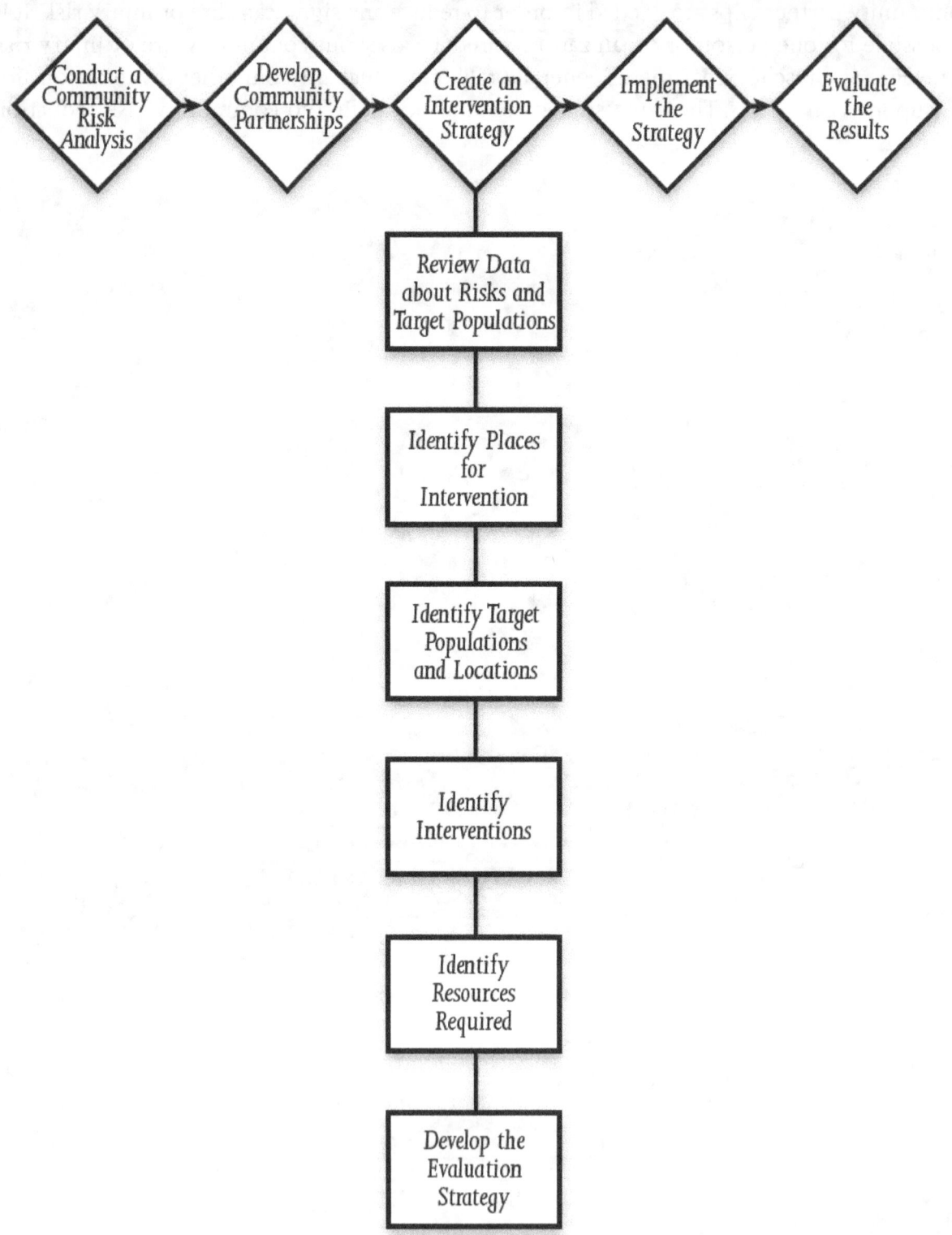

Chapter 3
Step 3: Create an Intervention Strategy

Introduction

What is an intervention strategy?

An intervention strategy is the beginning of the detailed work necessary for the development of a successful fire or life safety risk reduction program. The strategy should include what will be done, where it will be implemented, how the implementation will occur, and who will conduct the program once it is developed. It also should include an evaluation component that measures the effectiveness of the process and the program. Creating an intervention strategy requires a carefully thought-out plan of action developed through a group effort.

Why develop an intervention strategy?

Consider these questions:

*   Would a pro football coach send a team to the Super Bowl game without an action plan?

*   Would the commander-in-chief of the U.S. military send allied forces into battle without a plan of attack?

The answer to both questions: Of course not! Yet too many times an emergency service organization will try to address a community risk without an organized plan of action. Just as in the two examples above, money and lives are at stake.

Developing an intervention strategy is as important as conducting pre-emergency planning for high-hazard occupancies within a community. Taking the time needed to design a quality intervention strategy will help gain long-term risk reduction success.

An organized sequence of events should guide the development of the strategy. These events can be described as follows:

*   convene the community planning team and review the community profile and the problem statement;
*   identify places for intervention;
*   identify potential target populations and the physical locations for intervention opportunities;
*   identify specific interventions;
*   identify required resources; and
*   develop an evaluation plan.

The planning team

A community planning team is comprised of a core group of individuals who have a primary stake in the community risk issue or can offer key resources to the intervention process.

Safe City Example

Safe City selected the following planning team members:

- the community educator (and possibly select staff) from the Safe City Fire Department;
- members of the population affected by the cooking fires (older adults who reside at the housing units);
- the property owner (city housing authority) of the structures being damaged by the fires;
- community service providers (food and health services) who serve the older adult population;
- family members (children) of the older adult population;
- the president of the Housing Unit Tenant Association;
- media (TV, radio, newspaper) representatives;
- a representative from the local Commission on Aging;
- a professional experienced in market research; and
- a professional experienced in organizing and performing program evaluations (optional).

Review Data about Risks and Target Populations

The planning team must review the community profile and the problem statement so that everyone on the team has a factual description of the community and the problem. It is also important for the team to review the sources of the information used to develop the community profile and problem statement.

Develop a goal

A goal is a broad statement about the problem and the new condition the planning group would like to create.

Here is an example of a goal developed by the Safe City planning team in response to the cooking fire problem:

Safe City Example

Problem: A high number of cooking fires are occurring within the homes of older adults residing in city-owned housing units.

Goal: Decrease the number of cooking fires occurring within the homes of older adults residing in city-owned housing units by 25 percent in the first year, and then continue decreasing that number by an additional 25 percent in the 2 succeeding years.

Identify Places for Intervention

Review the analysis of the selected risk

Once the planning team has established a goal, the sequence of how and why a typical incident occurs should be revisited.

The planning team also should review the sequential analysis of a typical incident so that a broad range of prevention interventions can be considered. Consider the following possibilities:

- A person begins the cooking process on a stovetop. The heat is turned on high because the person has trouble seeing the control setting.

- A person forgets the stove is on and leaves the room to watch television or talk on the telephone.

- An unattended pan ignites on stovetop because the heat is set too high.

- The smoke alarm doesn't alert a person to a fire because the alarm is not in working condition.

- A person returns to the kitchen and performs inappropriate actions (for example, carrying a flaming pan from the home) because the person doesn't know the appropriate response to a stovetop fire.

Each of these scenarios has different implications for the most effective intervention.

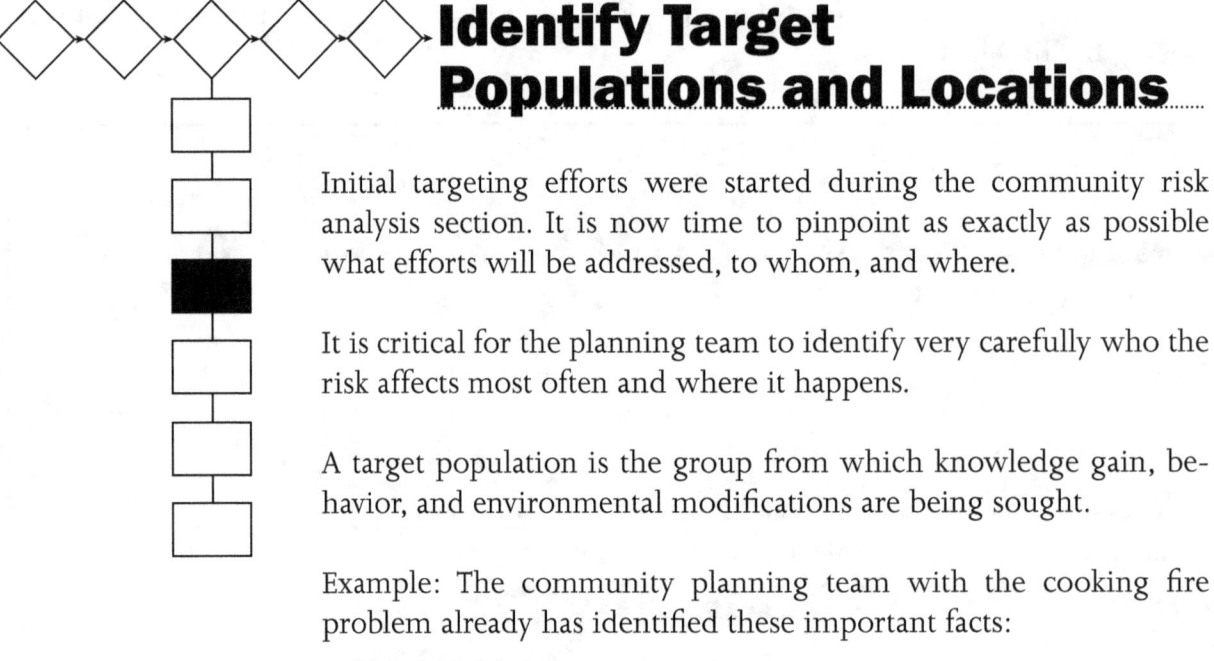

Identify Target Populations and Locations

Initial targeting efforts were started during the community risk analysis section. It is now time to pinpoint as exactly as possible what efforts will be addressed, to whom, and where.

It is critical for the planning team to identify very carefully who the risk affects most often and where it happens.

A target population is the group from which knowledge gain, behavior, and environmental modifications are being sought.

Example: The community planning team with the cooking fire problem already has identified these important facts:

- The cooking fire problem most often affects the older adult population residing independently within city-owned housing units.

- Knowledge gain, behavior, and environmental modifications must take place in these households if successful risk reduction is to occur.

- The housing units represent a logical place to direct initial efforts because support from both tenants and city officials is a realistic possibility.

Where the risk most often occurs is related to the population affected by the risk. In the above case the housing units represented a logical place to direct efforts. Another way to characterize where the risk occurs is to look at the geographic distribution of the risk in the city or county. In Chapter 1: Conduct a Community Risk Analysis the geographic component of the analysis process is discussed.

Once the team has identified all of the possible interventions, the primary population segment affected, and where the risk occurs most often, it is time to conduct market research.

Conducting market research

Market research is the process of learning about a population from that population. It involves direct communication with members of the target group. The goal is to gain insight into how risk reduction efforts should be designed and conducted.

Market research can be accomplished through direct interviews with people or by holding a focus group meeting. A focus group meeting includes people from the target population. These individuals offer insight into how their peers best receive and respond to information and requests. They also can provide important information about social and cultural issues specific to the group.

It is possible to enlist the services of a professional who specializes in market research. Perhaps this person or group may provide some services as an in-kind contribution to your organization.

Market research is a critical step in the risk reduction process that must not be overlooked. Do not assume you know everything about the target population. Seek information about a group's social, cultural, and political climate from members of the group.

If members of the proposed target population are not involved in the planning and solution process, the group may reject any intervention strategy that you attempt to implement. Don't take that chance!

Remember: people generally do not like having things done to them. Getting people to accept new ideas and make behavior changes can be a challenging process. Working with people from the target community from the beginning is vital to the success of the project.

Successful risk reduction efforts are more likely to be effective when the members of the target population:
- are aware of the problem;
- understand the problem and the factors that contribute to it;
- believe themselves, or their loved ones, to be personally at risk;
- believe that the risk is unacceptable and serious;
- understand that solutions to the problem exist;

- believe that changing their behavior will reduce the risk;
- believe that the benefits to change outweigh the barriers;
- believe that they are capable of making the expected behavioral change;
- are involved with the process from the beginning; and
- have an opportunity to provide input and suggestions.

> **Remember: the people from whom you are asking advice will offer valuable information about what may or may not work. Listen to them!**
>
> **Failure to heed advice from the community may result in the creation of a risk reduction strategy that is rejected by the target population.**

Let's look at a successful effort:

Safe City Example

Safe City conducted interviews with older adults at the housing units. A local market research professional also helped conduct a focus group meeting. This is what was learned about their older adult population through market research:

- Their older adult population was unaware of the existing cooking fire problem. Results of a survey conducted by the housing authority revealed that only 10 percent of residents could identify cooking as the leading cause of fire.

- Members of the group believed they were very safe from fire in their homes.

- Once they were made aware of the problem, the group believed they could be at risk. They were most concerned about those with mobility, vision, and hearing difficulties.

- Group members believed that their peers would help with prevention efforts, once they became aware of the problem and its potential solutions.

- The group respects the local emergency services as a credible source for risk and prevention information.

- When receiving information, group members prefer direct contact with people they know and trust. Examples: service providers, housing management, and fire department.

- The group enjoys and responds well to guest speakers at tenant association meetings.

- Upon reviewing the community profile, problem statement and sequential analysis, the group believes the risk is serious enough to pursue an organized prevention effort.

Once market research has been conducted and analyzed, the planning team must make some important decisions about what specific interventions will be pursued.

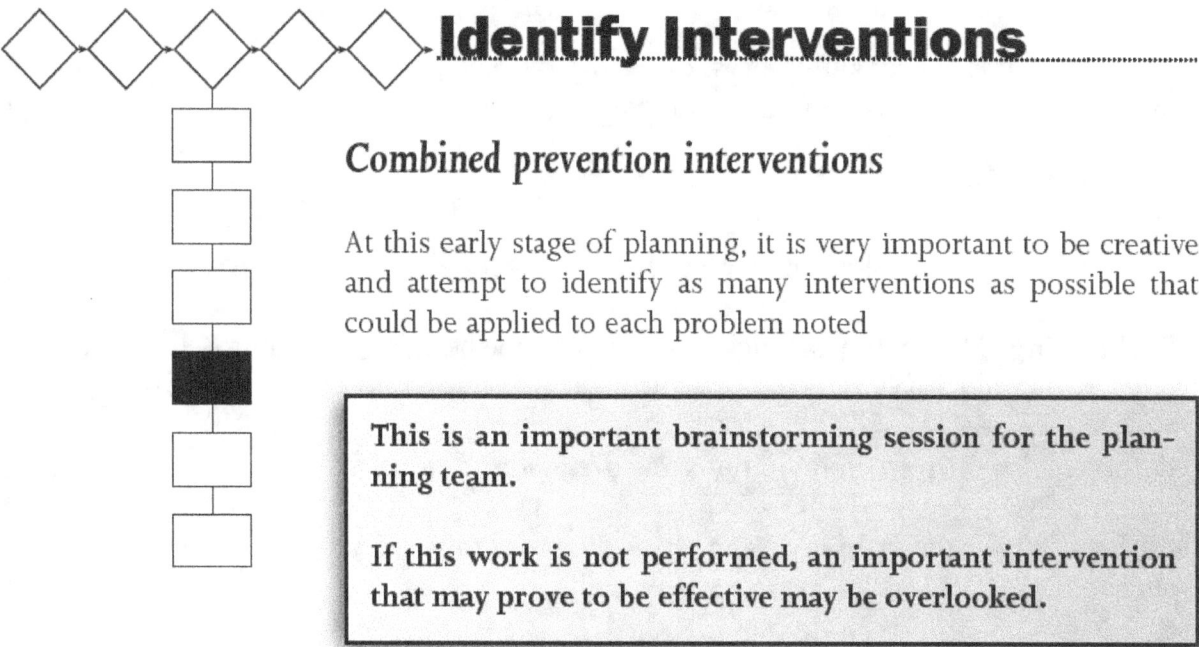

Identify Interventions

Combined prevention interventions

At this early stage of planning, it is very important to be creative and attempt to identify as many interventions as possible that could be applied to each problem noted

This is an important brainstorming session for the planning team.

If this work is not performed, an important intervention that may prove to be effective may be overlooked.

Most successful risk reduction efforts make use of combined prevention interventions. If the planning team with the cooking fire problem simply sends out a brochure about cooking fire prevention that urged older adults to be more careful, the cooking fire problem will not drop significantly. Combined prevention interventions include

- **Education:** Providing information (facts) about risk and prevention.

- **Engineering:** Using technology to create safer products or modifying the environment where the risk is occurring.

- **Enforcement:** Rules that require the use of a safety initiative.

> **Treat risk reduction as a science. Community partners and target populations must understand why risk occurs and how it can be prevented.**

Let's examine how to use combined interventions to address the cooking fire problem in Safe City for each possible scenario (the sequence of how and why a typical incident occurs).

Safe City Example

A person begins the cooking process on a stovetop. The heat is turned on high because the person has trouble seeing the control setting.

Intervention Ideas:

Education: Teach the older adults to cook with lower heat.

Engineering: Modify heat settings on stoves; provide better lighting at stovetops; make temperature dials easier to read.

Enforcement: Require modifications to stoves and stovetops.

A person forgets the stove is on and leaves the room to watch television or talk on the telephone.

Intervention Ideas:

Education: Educate the older adults to stay in the kitchen when using the stove.

Engineering: Relocate the TV and telephone to the kitchen area and/or install an alarm on the stove that sounds a warning that the device has been left unattended.

Enforcement: Require the TV and telephone to be near the kitchen area; require an alarm on stove; institute a penalty for people having an unattended cooking fire.

An unattended pan ignites on stovetop because the heat is set too high.

Intervention Ideas:

Education: Teach the older adults to stay in the kitchen and cook with lower heat.

Engineering: Create a stove that uses lower temperatures; install a stovetop fire suppression system; install fire-resistant building materials at the stove area.

Enforcement: Require installation of the above-listed equipment/materials.

The smoke alarm doesn't alert a person to a fire because the alarm is not in working condition.

Intervention Ideas:

Education: Reinforce the importance of working smoke alarms.

Engineering: Hard-wire smoke alarms and alarm systems into home electrical system; upgrade alarms to add enhanced audiovisual warnings.

Enforcement: Require updated smoke alarms and alarm systems; institute penalties for nonworking alarms.

A person returns to the kitchen and performs inappropriate actions (for example, carrying a flaming pan from the home) because the person doesn't know the appropriate response to a stovetop fire.

Intervention Ideas:

Education: Educate the older adults about what to do if a stovetop fire occurs.

Engineering: Install an automatic stovetop fire extinguishing system.

Enforcement: Require the installation of an automatic extinguishing system.

Do not cut corners with this step! Time and effort are required to consider and plan for the use of combined interventions. Used alone as an intervention, education may not produce the desired level of risk reduction.

After many intervention ideas have been identified, begin isolating specific and realistic areas within the sequence to intervene.

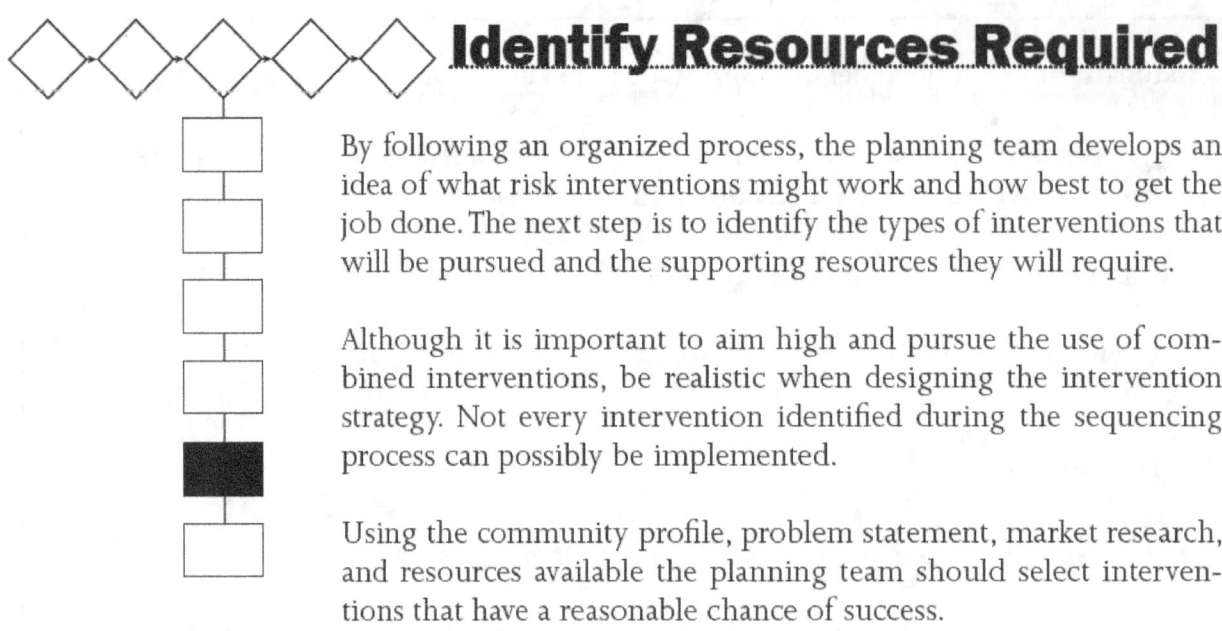

Identify Resources Required

By following an organized process, the planning team develops an idea of what risk interventions might work and how best to get the job done. The next step is to identify the types of interventions that will be pursued and the supporting resources they will require.

Although it is important to aim high and pursue the use of combined interventions, be realistic when designing the intervention strategy. Not every intervention identified during the sequencing process can possibly be implemented.

Using the community profile, problem statement, market research, and resources available the planning team should select interventions that have a reasonable chance of success.

Education intervention resources must be tailored to meet the needs of the local community. Materials must target the population and focus only on the key program messages. They must be created after your target injuries, population, intervention strategies, and implementation methods are established. Consult with your target audience and potential users concerning materials. They can tell you whether the materials do what you want them to do and how to revise them if necessary.

"Off the shelf" materials that have been developed by organizations such as the NFPA or the USFA can be useful in addressing local needs. It may be necessary to create local materials in order to address the specific needs of your community. If "off the shelf" materials do not have copyright restrictions then they can, in some cases, be modified to meet local needs.

As an example, let's look at the interventions that Safe City decided to pursue. Also included is the rationale for selecting each intervention and the resources that will be needed.

Safe City Example

Educational intervention: Provide awareness education about the cooking fire problem and proposed solutions to those who often interact with the older adult population residing in city-owned housing units.

Rationale: Those who frequently interact with the older adult population have the capability of delivering and reinforcing safety messages.

Resources needed: Consultation, collaboration, and support will be needed from all groups that are represented on the planning team.

Educational intervention: Provide awareness education about the cooking fire problem and proposed solutions to the older adult population residing in city-owned housing units.

Rationale: This is the population directly responsible for the occurrence of the problem. Their homes are catching fire! Behavior change is needed in this group.

Resources needed: Direct delivery of the message and political support of the proposed combined interventions.

Engineering intervention: Modification of the older adults' kitchens, including:

Short term:
- improve lighting near stove;
- install stove knobs with larger-print settings;
- install a telephone in the kitchen; and
- locate television within sight/hearing distance of stove;

Long term:
- install fire-resistant wall covering near stove area;
- install automatic fire extinguishing system over stovetop;
- upgrade smoke alarms and alarm system, including enhanced audiovisual warings; and
- replace aging stoves with new units that shut down prior to ignition temperature.

Rationale: Relying on education as a single intervention may not produce a significant drop in cooking fires, injuries, and property losses. Modifications to the kitchen environment and equipment should result in a safer food preparation area.

Resources needed: Employing engineering interventions and environmental modifications will take time, planning, and support. Funding, in-kind services, and political support must be sought from the groups represented on the planning team—in this case specifically, the city, housing authority, and tenants.

To sum it up: There must be a clear understanding of your community, its risks, the people, and the political climate to construct an effective intervention strategy.

The use of combined interventions often requires extensive political support. Obtaining such support will take time and effort but is well worth it. Now is the best time to start planning for the future.

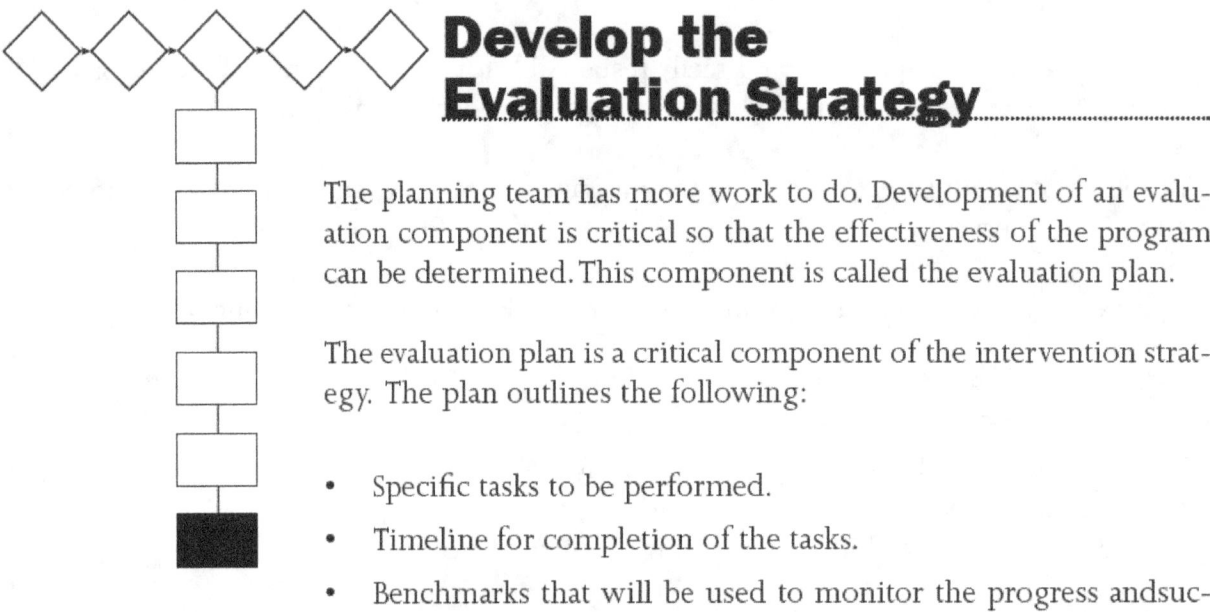

Develop the Evaluation Strategy

The planning team has more work to do. Development of an evaluation component is critical so that the effectiveness of the program can be determined. This component is called the evaluation plan.

The evaluation plan is a critical component of the intervention strategy. The plan outlines the following:

• Specific tasks to be performed.

• Timeline for completion of the tasks.

• Benchmarks that will be used to monitor the progress andsuccess of the intervention strategy.

Benchmarking is the comparison of various measures of effectiveness, performance, and cost to relevant standards, or benchmarks. The term comes originally from the field of surveying, where the benchmark is the topographical high point that serves as the point of reference against which all other elevations are measured. In the fire service the benchmark might be the fire rate for the State as a whole, or the national average, or the average of all those communities that local citizens and politicians see as equivalent to their jurisdiction.

One of the most useful benchmarking techniques is to calculate the average of a group of appropriate comparable communities. Therefore, the community will have a higher or lower fire rate, death rate, or injury rate than the average for communities of its size or composition. The comparison also can be made to the State or national average in the relevant categories. To measure that to a more exact standard, it is very easy to figure percentage comparisons to the average.

Why develop an evaluation plan?

Everyone has been involved with a project that failed because people or committees didn't follow through on what they promised to do.

Far too many times, a group starts out with the best intentions of completing a task only to have the effort fail because the group lacked an organized plan to keep things on track.

Adding an evaluation plan to the intervention strategy will help keep the risk reduction effort on track to reach the intended goal.

What is included in an evaluation plan?

An evaluation plan includes the problem statement, goal, and a series of objectives that support the goal. The objectives allow the evaluation of three levels of performance: outcome, impact, and process.

Here is an explanation of the three levels of evaluation:

- The outcome portion of the evaluation measures end results. Examples include changes in the occurrence of the specific risk issue, dollar loss, and the number of injuries and death.

- The impact portion of the evaluation measures knowledge, behavior, and environ-mental changes and also tracks changes in policy and legislation.

- The process portion of the evaluation measures the effectiveness of the program activity, the outreach effort, and the team members.

Why write objectives?

Having specific objectives will clearly identify

- what is going to be done;
- who is going to do it;
- when it is going to get done; and
- what the specific program benchmarks are.

Characteristics of well-written objectives

Well-written objectives have the following characteristics:

- The objectives are specific: they clearly identify what will be done, who will do it, and when.

- The objectives are measurable: they include a benchmark of what level of performance is being sought.

- The objectives are achievable: the measurement of desired performance is realistic based on local factors such as characteristics of the target population, social issues, and political climate.

- The objectives are consistent with the goal: they direct all activity and performance toward achieving the overall goal.

How do you create an evaluation plan?

The first and most important step in developing an evaluation plan is to understand that creating the evaluation plan is relatively simple. The hardest work (the intervention strategy) already has been completed.

The evaluation plan is created based on the intervention strategy. It simply involves finding out where the community is statistically (baseline) and plotting where the community wants to be (benchmark). In the case of outcome and impact objectives, a baseline measurement is required, so that existing conditions can be compared with future performance.

Let's look at the interventions the Safe City, USA, team decided to pursue, and look at the evaluation plan it prepared.

Safe City Example

Educational interventions:

- Provide awareness education about the cooking fire problem and proposed solutions to those who often interact with the older adult population residing in city-owned housing units.

- Provide awareness education about the cooking fire problem and proposed solutions to the older adult population residing in city-owned housing units.

Engineering interventions: Make short-term and long-term modifications in older adults' kitchens.

Enforcement interventions: Require the city to proceed with the identified engineering interventions and environmental modifications.

Here's an evaluation plan for the intervention strategies selected by the Safe City planning team.

Evaluation Plan

Problem: A high number of cooking fires occur within the homes of older adults residing in city-owned housing units.

Goal: A significant decrease in the number of cooking fires occurring within the homes of older adults residing in city-owned housing.

Outcome Objective: Within 4 years, the number of cooking fires that occur in homes

---

of older adults residing in city-owned housing units will be reduced by more than 50 percent.

Recall that Safe City Fire Department learned in its community risk analysis that one half of its yearly average of 40 cooking fires occurred in older-adult housing units. The baseline for measurement is therefore 20 cooking fires per year.

Note: An evaluation plan usually has one outcome objective. The reason: Outcome measures the bottom line–the degree to which the program has reduced occurrence of risk, dollar loss, and injuries or deaths. Outcome objectives are usually expressed 3 to 5 years in advance. Impact objectives can be expressed within 1 to 3 years. Process objectives can be expressed within a timeframe of 6 months to 2 years.

Method of evaluation: Annual response statistics.

Group responsible: Fire department staff and the planning team, led by a community educator.

The following impact objective and four-process objectives support the educational interventions outlined in the intervention strategy:

Impact Objective: Within 2 years, 70 percent of older adults living in city-owned housing will identify cooking as the leading fire cause and state appropriate prevention/reaction strategies

The baseline for measurement is 10 percent. Recall that the planning team asked the housing authority to survey older adults and seek opinions on the leading cause of fire. Only 10 percent of residents identified cooking as the leading cause.

Method of evaluation: The results of a second housing authority survey conducted after completion of educational interventions.

Group responsible: Fire department and housing authority staff; planning team members.

Process Objective 1: By the end of 6 months, 100 percent of the planning team members will have conducted cooking fire awareness training with their organization's staff.

Method of evaluation: The organization's record of training.

Group responsible: Planning team members.

Process Objective 2: By the end of year one, 80 percent of the organizations represented by planning team members will be conducting cooking fire awareness efforts among the older adult population residing in city-owned housing units.

Method of evaluation: The organization's record of staff activity.

Group responsible: Planning team members.

Process Objective 3: By the end of year one, the fire department will have presented cooking fire awareness programs at all city-owned older-adult housing units.

Method of evaluation: Fire department staff activity records.

Group responsible: Fire department staff, led by a community educator.

Process Objective 4: By the end of 18 months, the fire department (supported by members of the planning team) will have made direct in-home contact with 50 percent of older adults living in city-owned housing to discuss cooking fire awareness.

Method of evaluation: Fire department record of visitations.

Group responsible: Community educator, fire department staff, and planning team members.

Why this intervention and evaluation strategy will be successful

Safe City has set itself up for a successful risk reduction outcome. Its intervention and evaluation strategy will lead to success because:

- The strategy is based on facts. The community profile and problem statement are based on information obtained through a close examination of the local community.

- The strategy is specific. The strategy focuses on the leading risk issue (cooking) and population (older adults) most affected by the risk.

- The strategy is realistic. The strategy focuses on older adults living in city-owned housing. It is realistic because the core group of primary stakeholders is relatively small.

This strategy is a good starting point for gaining the political support needed to promote and use combined prevention interventions.

- The strategy is community-based. The fire department, planning team, older adults, and housing authority will take part in the problem-solving process.

- The strategy will be implemented over time and will receive periodic evaluations.

- Evaluation will be an ongoing process so that activities can be monitored and adjusted as needed.

Develop an evaluation plan with assistance from the planning team. Base the plan on information from the community profile, problem statement, and market research.

Summary

Creating an intervention strategy involves detailed work and a team approach in order to develop a successful risk reduction effort. The strategy must include what will be done, where it will be done, how the implementation will occur, and who will conduct the program. An evaluation component measures the effectiveness of the process and the program. Creating an intervention strategy requires a carefully thought-out plan of action based on a review of the risk data, the target population, locations, and available resources.

Step 4: Implement the Strategy

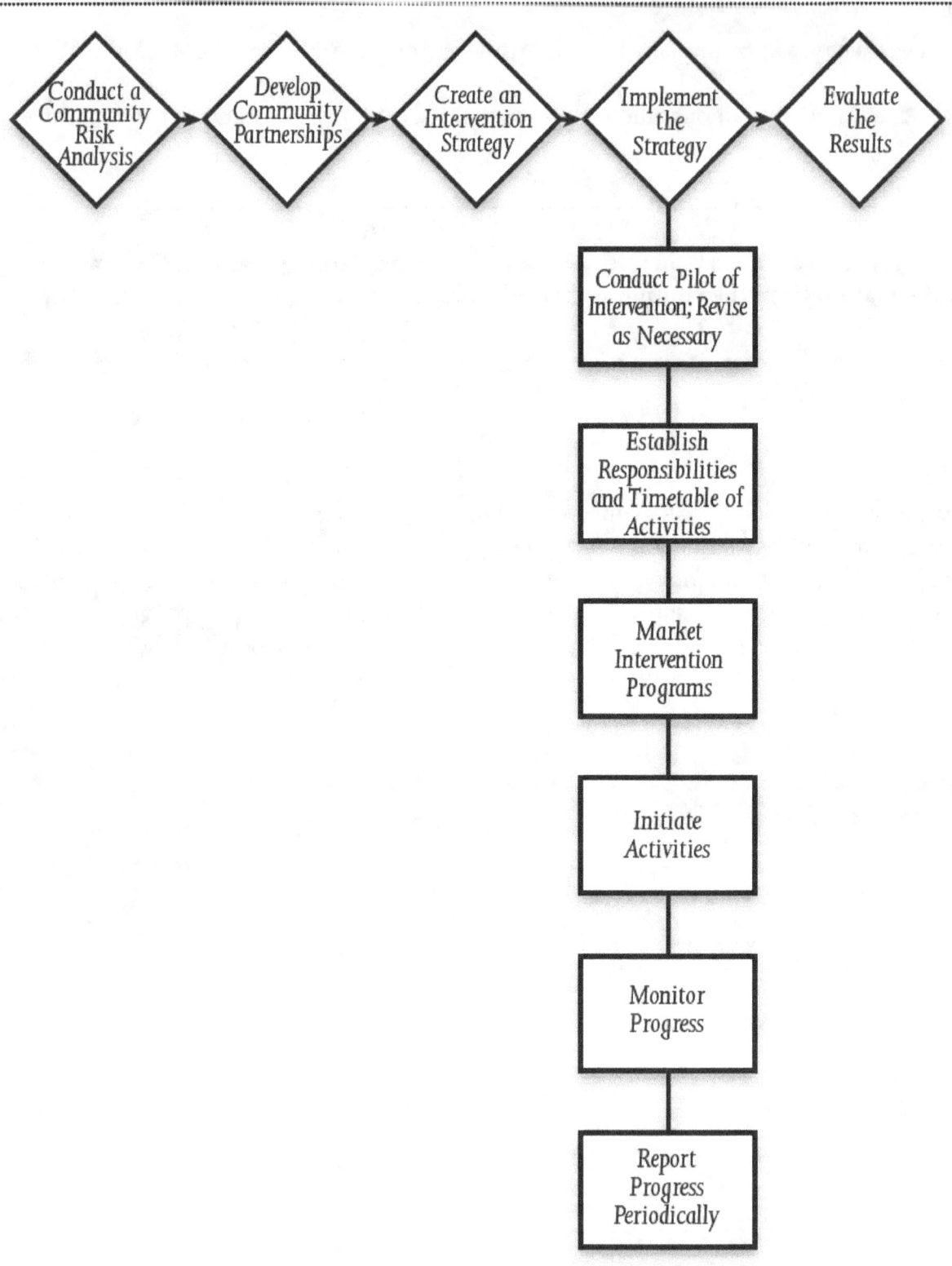

Conduct a Community Risk Analysis → Develop Community Partnerships → Create an Intervention Strategy → Implement the Strategy → Evaluate the Results

Conduct Pilot of Intervention; Revise as Necessary

Establish Responsibilities and Timetable of Activities

Market Intervention Programs

Initiate Activities

Monitor Progress

Report Progress Periodically

Chapter 4
Step 4: Implement the Strategy

Introduction

Implementing the strategy puts the plan into action

Implementing the strategy involves testing the interventions and then putting the plan into action in the community. Sometimes modifications are made to the program as the result of a pilot implementation. In a real sense, the implementation step is where "the rubber meets the road."

Think about the interventions developed by the Safe City, USA, Fire Department for the cooking fire problem. The interventions included an awareness education program for the older adults and for those whose provide services to the older adults. Other interventions included modifications of the kitchens of older adults. The implementation plan identifies the steps to take to bring those interventions into the community.

The implementation plan outlines the steps for implementing the program in the community and provides the following details:

- How the program will be implemented, including when, how long, where, etc.

- It identifies the roles and responsibilities of the program implementation team.

- The process for the pilot test implementation.

- Provisions for making any modifications to the program based on the results of the pilot implementation.

- A predelivery checklist identifying tasks that must be done prior to the implementation.

- A description of potential implementation problems and contingencies.

An action plan is a step-by-step outline of work that needs to be done in order to meet the stated objective. Each objective requires its own action plan. An Action Planning Chart is a useful tool that can be used in planning the implementation as well as in monitoring progress.

Action Plan   Date:_____

Goal: **Objective:**

Step#	Action	To Be Completed By	Person Responsible	Resources Needed	Date Completed

Figure 11: *Action Planning Chart.*

The actions in the implementation step:

1. Conduct pilot of intervention; revise as necessary.

2. Establish responsibilities and timetable of activities.

3. Market intervention program.

4. Initiate activities.

5. Monitor progress regularly.

6. Report progress periodically.

Develop the implementation plan by working together with the community team and representatives from the target audience. The implementation plan simply identifies the best approach for putting the interventions into action.

There is a temptation just to put the program into action without an implementation plan. Take the time to develop a step-by-step implementation plan. An essential element of the implementation plan is the identification of the roles and responsibilities of each team member. One of the most common problem areas during implementation is the confusion surrounding who should be doing what, and when it should be done. The implementation plan clearly spells out individual and group tasks, authority, and overall responsibility.

Some people actually will deliver the presentations. Some will assist with scheduling the presentations and making connections with the older adults. Others will be involved in the evaluation process. The implementation plan should describe what is to be accomplished by each task, and when each task will be completed.

Match people with tasks that suit them well, and that they are prepared to complete.

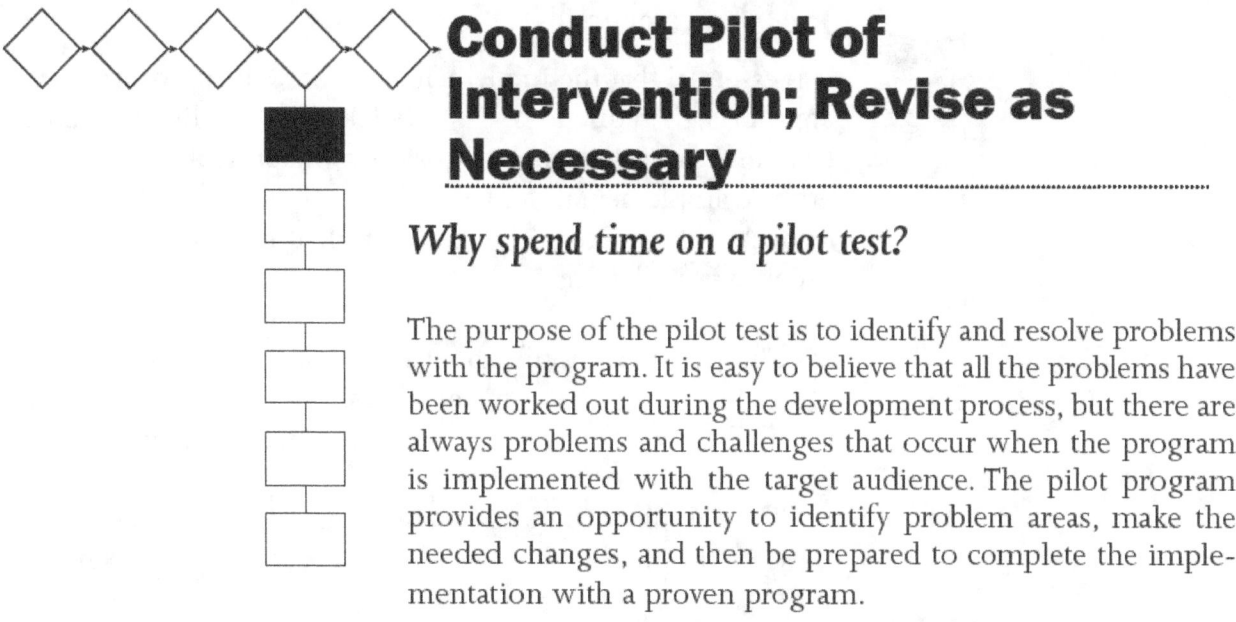

Conduct Pilot of Intervention; Revise as Necessary

Why spend time on a pilot test?

The purpose of the pilot test is to identify and resolve problems with the program. It is easy to believe that all the problems have been worked out during the development process, but there are always problems and challenges that occur when the program is implemented with the target audience. The pilot program provides an opportunity to identify problem areas, make the needed changes, and then be prepared to complete the implementation with a proven program.

Evaluating of the pilot program involves the evaluation of two different areas: the program results and the pilot implementation process. The implementation process includes how to gather data, the methods to be used to interpret the data, and the people who will do the evaluation. The information gathered as part of this process is used to modify the program prior to full implementation.

The pilot program offers an opportunity to conduct the evaluation of the program on a small scale. This allows more interaction with the target audience and with those conducting the program. The evaluators should provide an objective review of the program. Remember, evaluation is a key part of every program. If it is worth doing it is worth evaluating.

The success of the overall program may depend on the effectiveness of the pilot implementation. A representative of the target audience should be involved throughout the pilot test to provide feedback directly to the development team.

Modifying the program

The evaluation of the pilot program should provide information on what worked and what didn't work. Be careful not to make drastic changes without solid justification. Also, keep the target audience representative closely involved in the modifications. Finally, ensure the modifications do not take away from the strengths of the program.

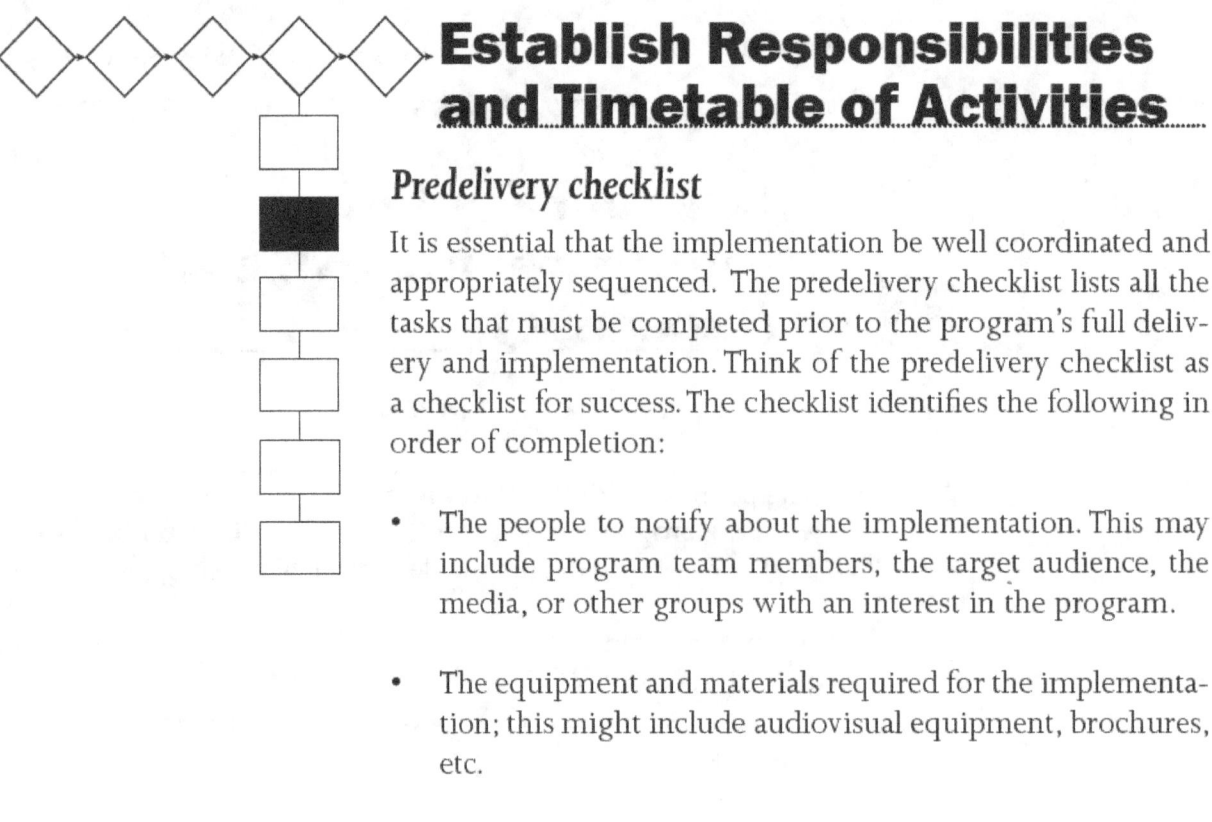

Establish Responsibilities and Timetable of Activities

Predelivery checklist

It is essential that the implementation be well coordinated and appropriately sequenced. The predelivery checklist lists all the tasks that must be completed prior to the program's full delivery and implementation. Think of the predelivery checklist as a checklist for success. The checklist identifies the following in order of completion:

- The people to notify about the implementation. This may include program team members, the target audience, the media, or other groups with an interest in the program.

- The equipment and materials required for the implementation; this might include audiovisual equipment, brochures, etc.

- Appointments and meetings that must be scheduled including presentations to the target audience.

- Transportation and other support requirements.

The checklist is a road map for the tasks that must be completed in order to implement the program. However, the checklist isn't foolproof. Expect additional needs to arise in the process.

The Safe City predelivery checklist below provides an example of some of the tasks that must be completed as part of the implementation of this intervention.

Safe City Example

Predelivery Checklist

☐ Contact managers of the city-owned housing units to schedule a 1-hour presentation to the residents.

☐ Recruit personnel to deliver the presentations.

☐ Send the handout materials and videotapes to the presenters.

☐ Contact the media with a news release about the program.

Implementation problems

Seldom is there a community education initiative that doesn't experience problems during implementation. However, some common and expected problems can be identified and prepared for with a contingency plan.

To address potential problems, bring the development team together with the representative(s) of the target audience and brainstorm potential problems and appropriate solutions. It is especially helpful to involve educators and community leaders who have been involved in similar programs in the brainstorming process. Their experience is invaluable in identifying potential problems.

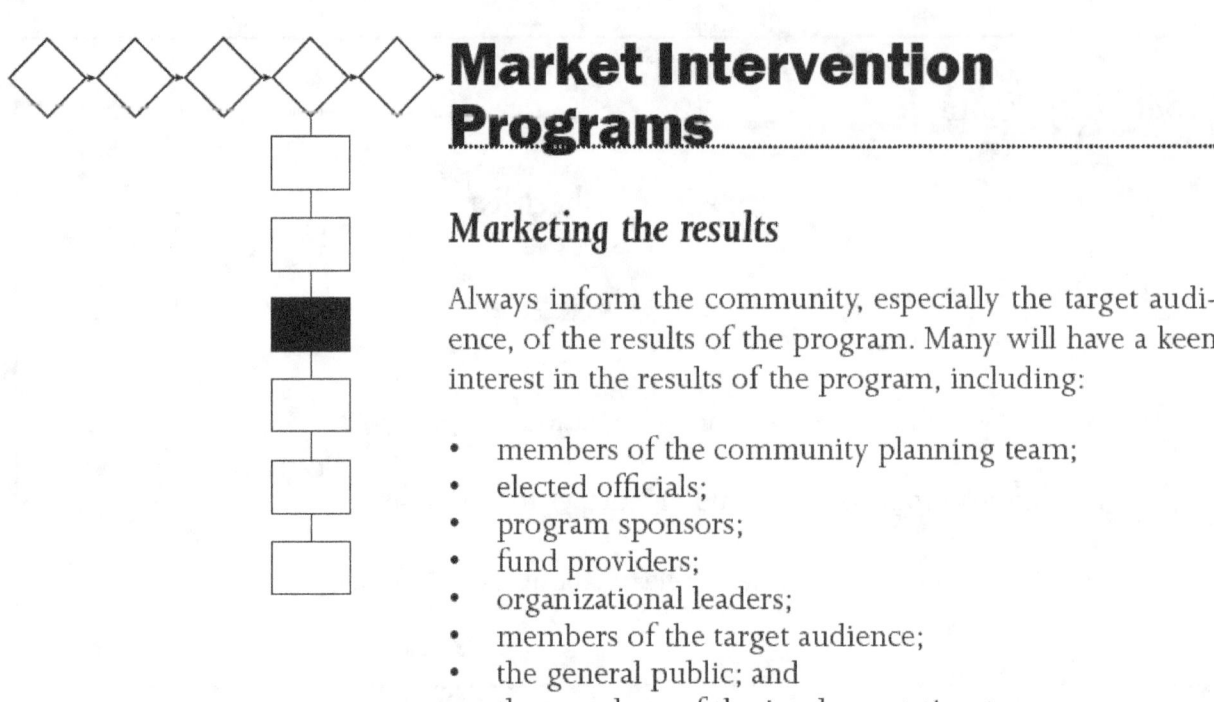

Market Intervention Programs

Marketing the results

Always inform the community, especially the target audience, of the results of the program. Many will have a keen interest in the results of the program, including:

- members of the community planning team;
- elected officials;
- program sponsors;
- fund providers;
- organizational leaders;
- members of the target audience;
- the general public; and
- the members of the implementation team.

The purpose of marketing the results of the program is to inform people about the effectiveness of the program in reducing fires, injuries, etc. Let the community know the program works! Future programs may depend on how well you market the results of the current program.

What kind of information should be communicated?

The information communicated to the community may include

- the educational gain in the target audience;
- the details of program activities, including the number of presentations conducted;
- anecdotes of those involved in the program, including members of the target audience; and
- report on decrease in the number of fires, injuries, etc.

It is easy to disseminate information that doesn't mean a thing to the public.

Always use terms that are easy to understand, and make statistics simple and easy to grasp. Have someone who has not been involved look at the materials to see if they make sense.

How to get information out

Some of the more common methods of disseminating information are provided below.

- **Local media**—Using local media is the best way to reach the general public. This includes news stories, articles, and features. The local media can be reached through phone calls, news releases, or news conferences.

- **Newsletters**—Newsletters generally have specific audiences. A story in a newsletter may reach a target audience that otherwise would be difficult to reach.

- **Direct mailing**—If the program involves a specific geographic area, a direct mailing may be effective. Address lists often are available through the county clerk's office; the local Chamber of Commerce; utility departments; etc.

- **Meetings**—To reach members of the coalition and development team, meetings are the fastest and most effective way to disseminate the information.

Consider recruiting a marketing professional to assist with marketing. This person will have experience with reaching specific groups with information.

Team recognition

Successful programs are the result of hard work by people who are committed to making things better in their communities and who are willing to make a personal sacrifice to bring that change to reality.

You should recognize anyone who has been involved in the development and implementation of the program. Of course, those who contribute the most should receive the highest recognition. There are many ways to recognize the efforts of the team; some of those methods are listed below.

- certificates of appreciation;
- gift certificates to local businesses;
- items that can be used at home or work;
- letters of appreciation by local officials; and
- appreciation dinners or other events.

Regardless of the recognition method selected, ensure that the recognition is sincere and meaningful. Recognition of the team members says, "Thank you for a job well done." It sends the message that the sacrifice and hard work had meaning to you and to the community. It tells people that they are important and the mission of the program is important. Without recognition people may feel taken advantage of. The importance of their efforts may be minimized in their minds and in the mind of the community.

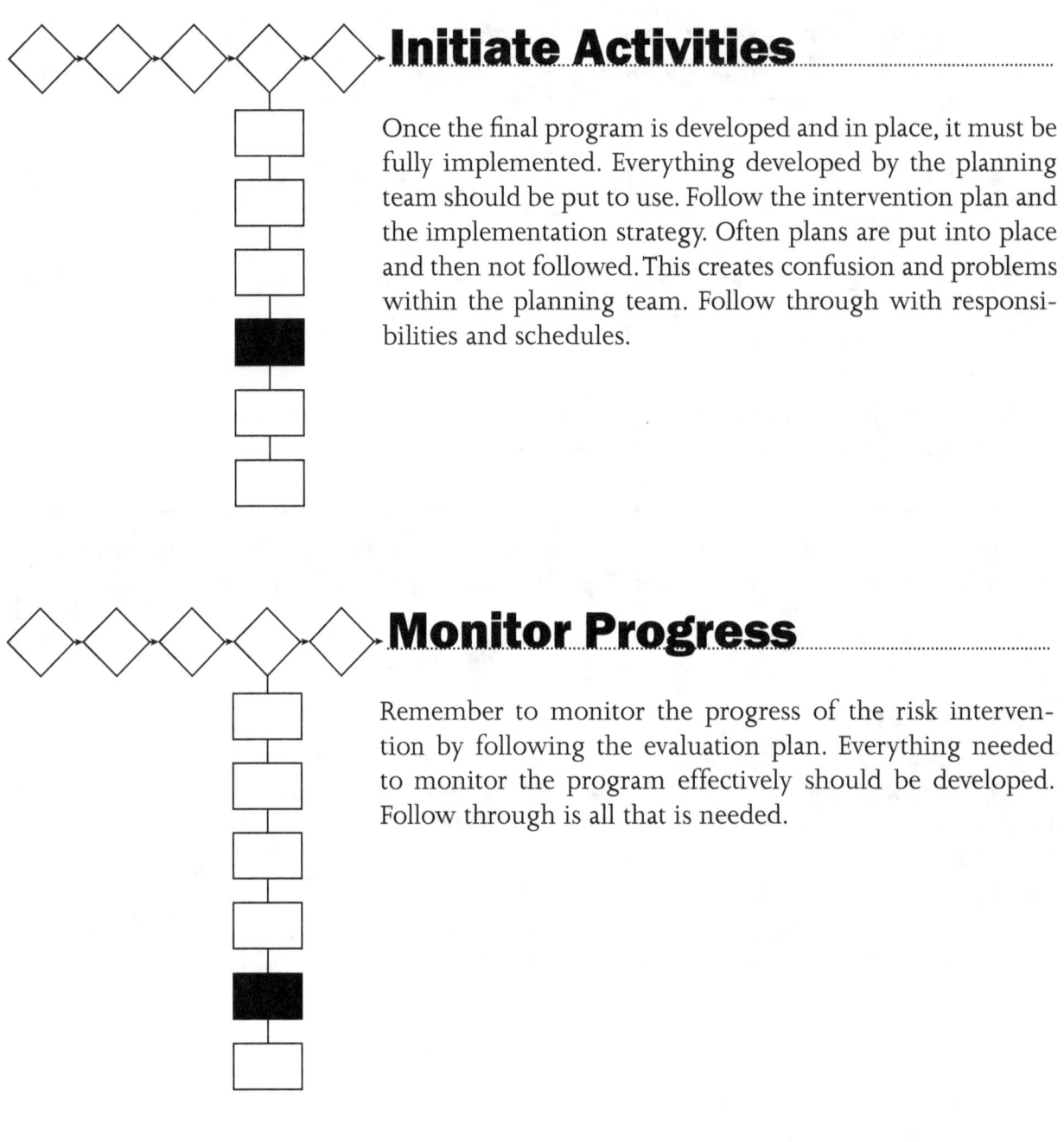

Initiate Activities

Once the final program is developed and in place, it must be fully implemented. Everything developed by the planning team should be put to use. Follow the intervention plan and the implementation strategy. Often plans are put into place and then not followed. This creates confusion and problems within the planning team. Follow through with responsibilities and schedules.

Monitor Progress

Remember to monitor the progress of the risk intervention by following the evaluation plan. Everything needed to monitor the program effectively should be developed. Follow through is all that is needed.

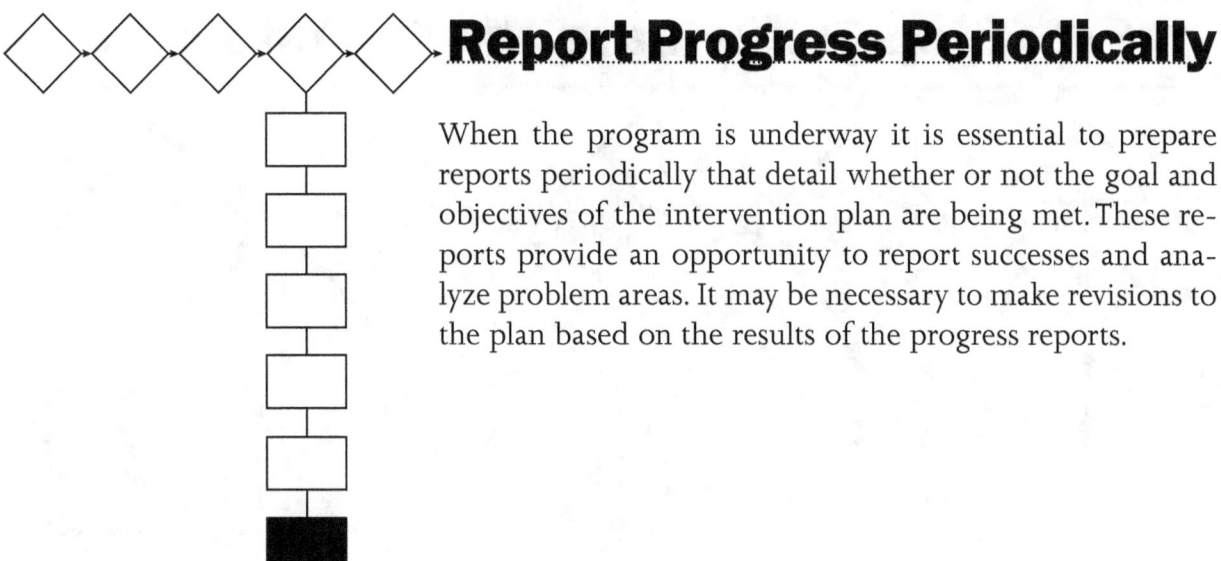

Report Progress Periodically

When the program is underway it is essential to prepare reports periodically that detail whether or not the goal and objectives of the intervention plan are being met. These reports provide an opportunity to report successes and analyze problem areas. It may be necessary to make revisions to the plan based on the results of the progress reports.

Summary

Implementing the strategy involves pilot testing the interventions and then putting the plan into action in the community. Modifications are sometimes made to the program. Marketing the program is an important part of implementation that should not be overlooked. Monitoring the progress of the program implementation and preparing regular progress reports also are important.

Step 5: Evaluate the Results

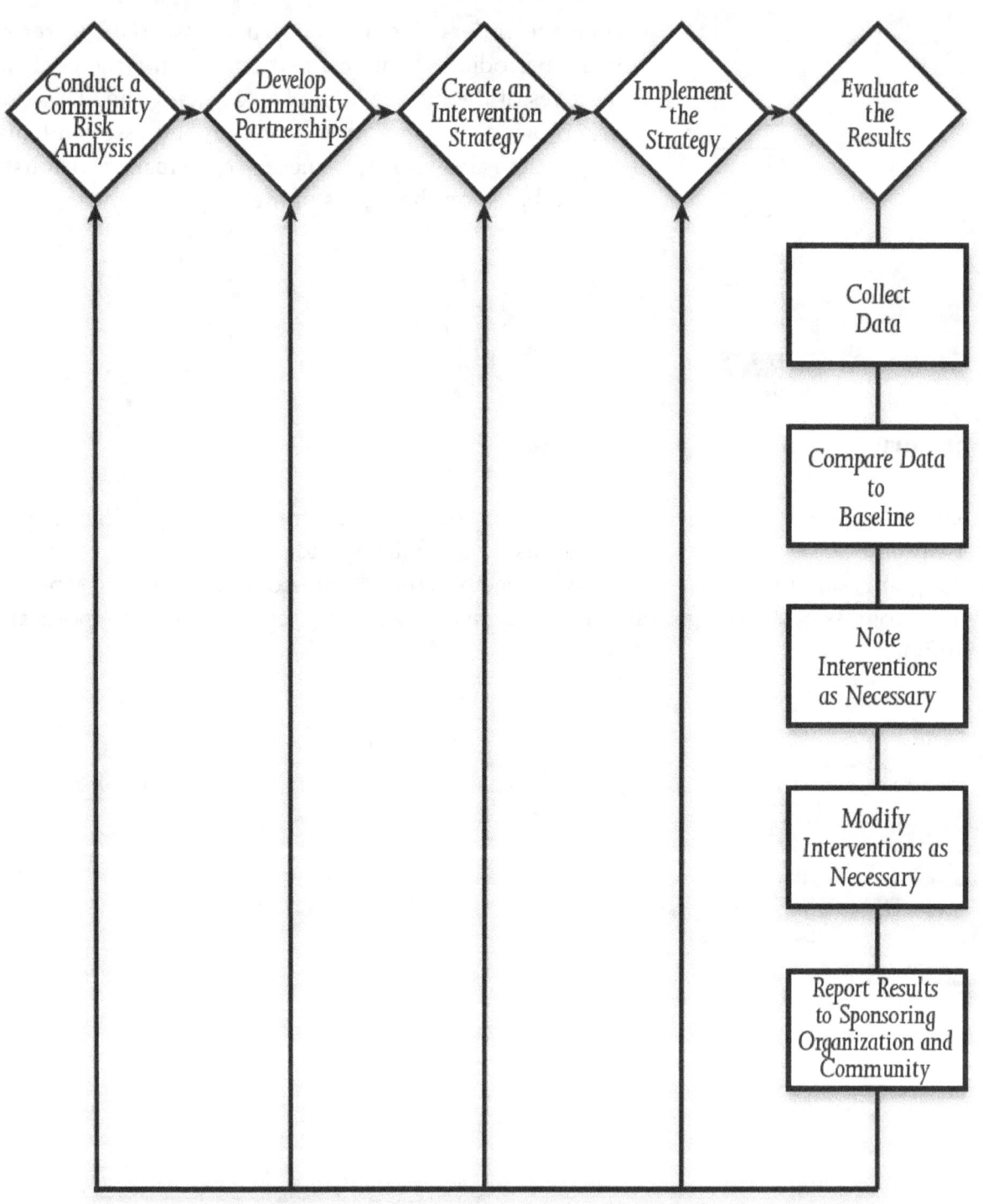

Chapter 5
Step 5: Evaluate the Results

Introduction

Why evaluate the risk reduction process?

The primary goal of the evaluation process is to demonstrate that risk reduction efforts are reaching target populations, have the planned impact, and are demonstrably reducing loss.

With so much at stake, why guess if risk reduction efforts are being successful? Prove it!

Some organizations skip evaluations

While there are many reasons why an organization may not evaluate a risk reduction effort, here are the three big ones:

1. Fear of working with statistics.

   Many people believe that evaluation involves complex mathematical formulas. Not so. It's all about identifying existing conditions, determining future desired performance levels, making comparisons, and reaching conclusions. The use of simple math (addition, subtraction, multiplication, and division) will get the job done.

2. Fear that a good evaluation may identify shortcomings in program efforts.

3. Lack of knowledge about evaluation.

Many organizations are simply not sure how to perform evaluation. This situation can be handled rather easily by seeking help. Consider these sources:

- USFA;
- NFPA;
- other emergency service organizations that have experience with evaluation;
- teachers and school officials;
- public health officials;
- counselors and mental health officials;
- planning and development officials;
- Census Bureau; and
- community colleges and universities.

How can evaluation prove that the risk reduction process is on track?

A good evaluation process can prove that an intervention program is effective by validating that the goal of the intervention is being met by citing objective data. Evaluation is all about measuring performance.

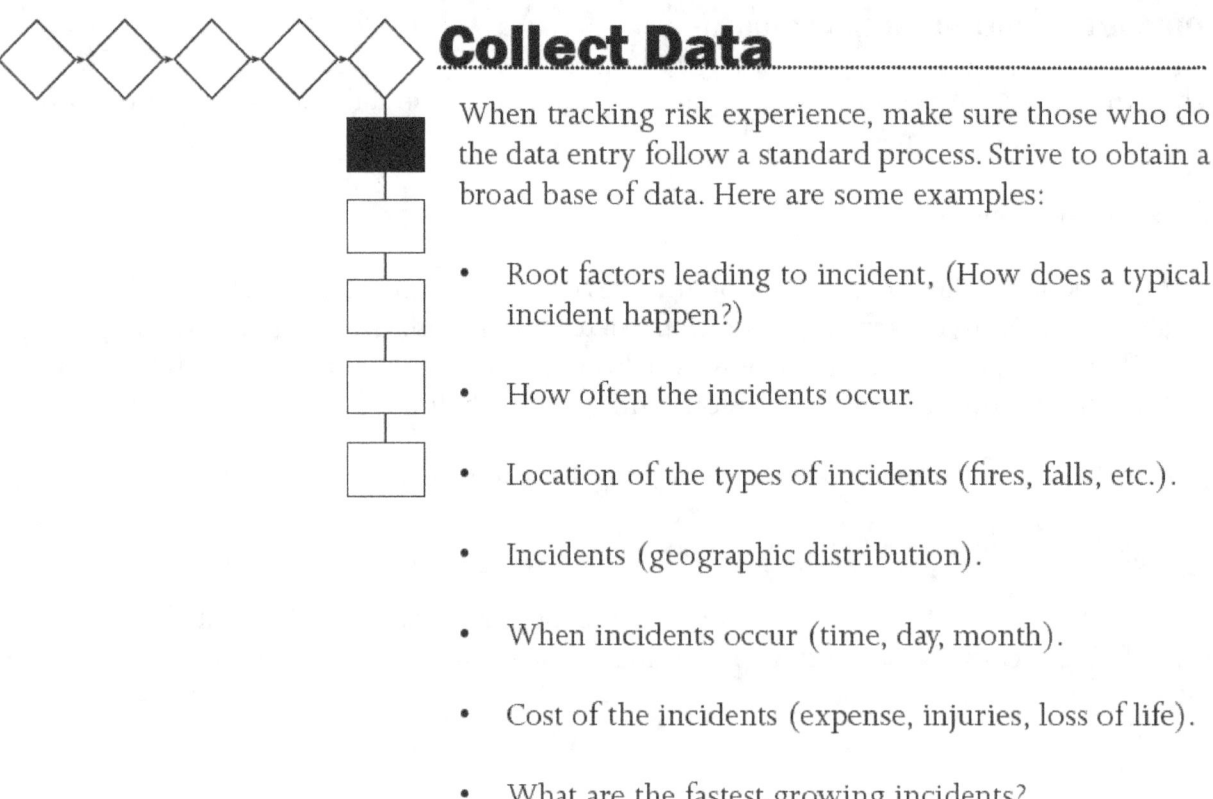

Collect Data

When tracking risk experience, make sure those who do the data entry follow a standard process. Strive to obtain a broad base of data. Here are some examples:

• Root factors leading to incident, (How does a typical incident happen?)

• How often the incidents occur.

• Location of the types of incidents (fires, falls, etc.).

• Incidents (geographic distribution).

• When incidents occur (time, day, month).

• Cost of the incidents (expense, injuries, loss of life).

• What are the fastest growing incidents?

When using a survey instrument or a questionnaire, make sure it is designed to seek information in an objective manner.

Example—You want to find out if people know the leading cause of fire in the community.

Ask—What do you believe is the leading cause of fire in the community?

Instead of—Do you think unattended cooking is the leading cause of fire in the community?

> **Asking a question that leads a person to a specific response is likely to produce a biased answer.**

Evaluation drives the risk reduction process. The most effective risk reduction efforts are those led by people willing to modify program strategies based on the results of ongoing evaluation.

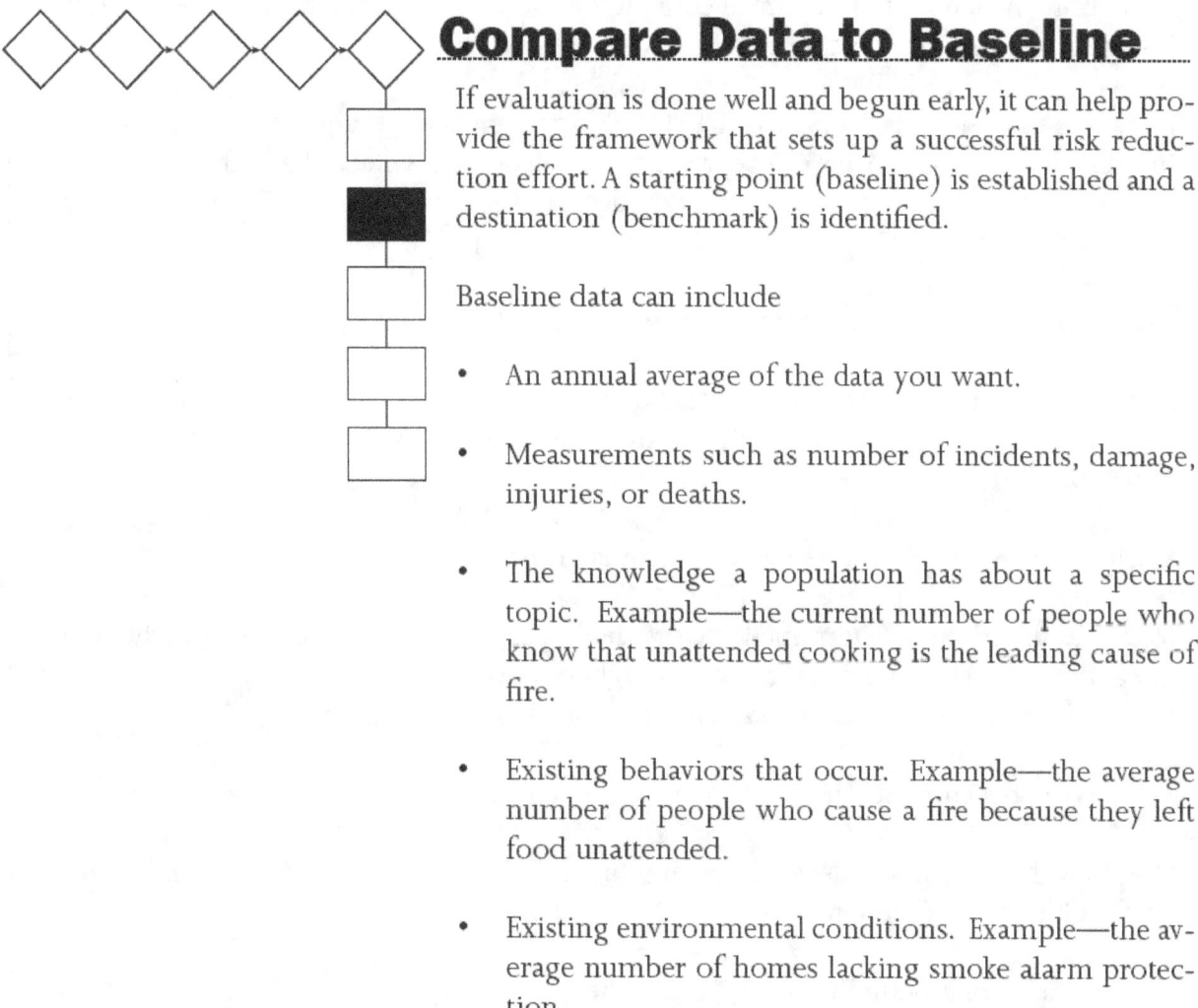

Compare Data to Baseline

If evaluation is done well and begun early, it can help provide the framework that sets up a successful risk reduction effort. A starting point (baseline) is established and a destination (benchmark) is identified.

Baseline data can include

- An annual average of the data you want.

- Measurements such as number of incidents, damage, injuries, or deaths.

- The knowledge a population has about a specific topic. Example—the current number of people who know that unattended cooking is the leading cause of fire.

- Existing behaviors that occur. Example—the average number of people who cause a fire because they left food unattended.

- Existing environmental conditions. Example—the average number of homes lacking smoke alarm protection.

The benchmark is the desired level of change. Let's look at an example from Safe City that identifies baseline and future data needed to make conclusions regarding the success of the risk reduction effort. There is also an examination one of their objectives for each level of performance—outcome, impact, and process.

Safe City Example

Cooking Fire Problem in Safe City, USA

Outcome Objective: By the end of year one there will be a 50 percent reduction in the number of cooking fires that occur in homes of older adults residing in city-owned housing units.

What's the baseline? 20 cooking fires per year.

What is the benchmark? Cut annual cooking fire occurrence in the homes of older adults living in housing to 10 per year.

How was the benchmark decided? The fire department is targeting the older adult housing unit because they believe it represents a realistic area where risk reduction will be successful. The fire department wants to decrease cooking fires by 50 percent over a 2-year period (20/2=10).

How is progress being monitored? Review of annual response statistics.

Impact Objective: By the end of year one, 70 percent of older adults living in city-owned housing will identify cooking as the leading fire cause and state appropriate prevention/reaction strategies.

What's the baseline? Baseline measurement is 10 percent.

How was the baseline determined? Before the impact objective was developed, the housing authority surveyed older adults and sought opinions on the leading cause of fire. One hundred responses were obtained. Only 10 residents said cooking.

What is the benchmark? The benchmark is 70 percent.

How was the benchmark decided? During the initial survey, only 10 percent of the older citizens knew the correct answer.

How can progress be monitored? During the next survey (to be done after the educa-

---

tional interventions) 70 out of 100 older adults must identify cooking as the leading fire cause.

Process Objective: By the end of 6 months, the fire department (supported by members of the planning team) will have made direct in-home contact with 50 percent of older adults living in city-owned housing to discuss cooking fire awareness.

What's the baseline? In this case, the baseline is zero because the fire department has not performed any previous door-to-door visitation at the older adult housing unit.

What is the benchmark? The benchmark is 50 percent.

How was the benchmark obtained? The fire department will visit 100 homes within the housing unit. They project that a successful contact (person at home) will take place on 50 visits.

How can progress be monitored? A record of visitations.

Important facts to remember about evaluation

Evaluation must span the risk reduction process. It begins with analyzing and planning to address risk. It continues through implementation. Modifications to program activity are made according to evaluation results.

By creating and following a well-written evaluation plan, the specific progress of the risk reduction effort can be monitored and adapted according to need.

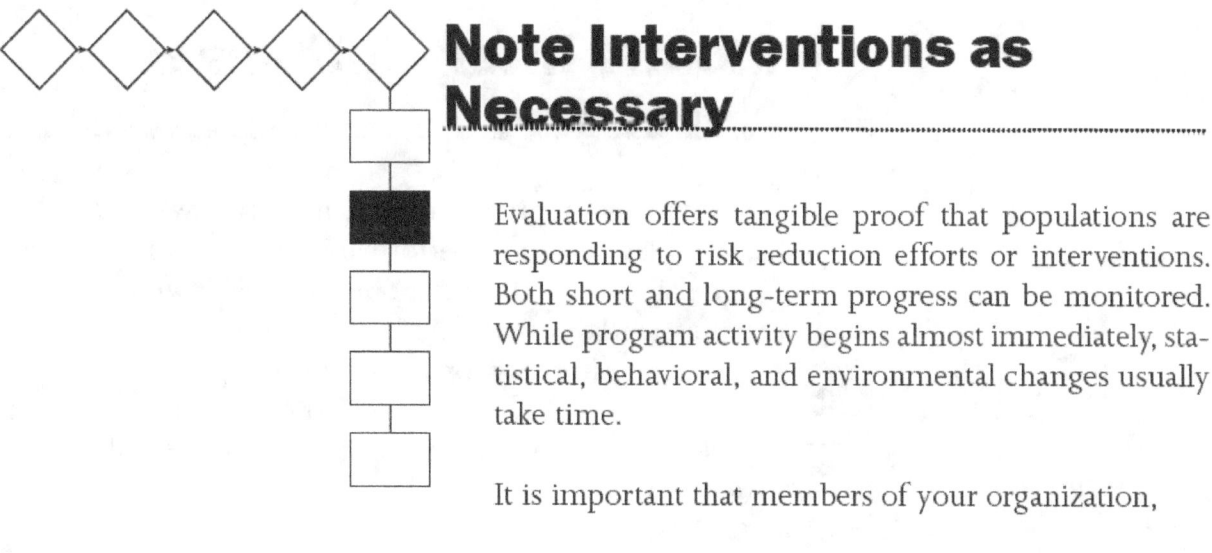

Note Interventions as Necessary

Evaluation offers tangible proof that populations are responding to risk reduction efforts or interventions. Both short and long-term progress can be monitored. While program activity begins almost immediately, statistical, behavioral, and environmental changes usually take time.

It is important that members of your organization,

planning team, and community be able to see steady progress being made toward reaching the set goal.

Often, the myth that prevention cannot be measured has prevented organizations from attempting to conduct quality evaluation efforts. Prevention efforts can be measured quite well so long as the organization makes a commitment to do so in the three levels outlined in this text.

Outcome and impact evaluation depend upon measuring before and after the intervention. Compare the baseline data to data collected during and after risk reduction efforts have been done. The data then are compared to the benchmark set in the evaluation plan.

Safe City Example

Safe City Performance Measurements

Knowledge: The percent of a population who know the leading cause of fire in the community.

Behavior Change: The percent of a population that stay in the kitchen when cooking food.

Environmental Change: The percent of homes that have a working smoke alarm.

Legislative Change: A law is enacted that requires smoke alarms in all homes.

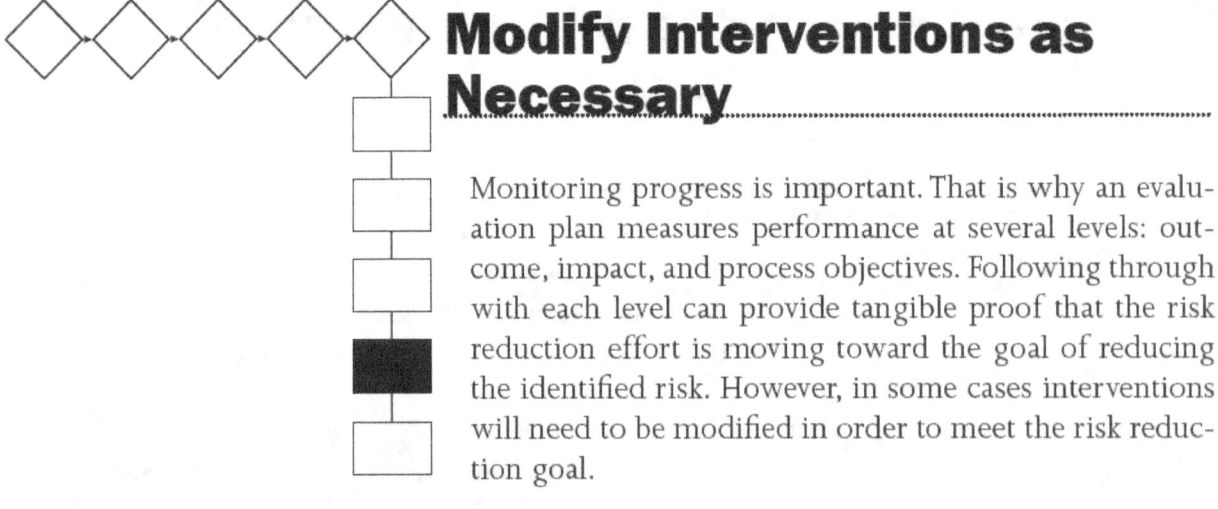

Modify Interventions as Necessary

Monitoring progress is important. That is why an evaluation plan measures performance at several levels: outcome, impact, and process objectives. Following through with each level can provide tangible proof that the risk reduction effort is moving toward the goal of reducing the identified risk. However, in some cases interventions will need to be modified in order to meet the risk reduction goal.

Evaluation takes time

Outcome evaluation is long term. While annual statistics are used to monitor changes in risk, it may take 3 to 5 years (or longer) to realize major progress.

The results of impact evaluation may be seen a bit more quickly, but not overnight! While changes in knowledge can be realized over a relatively short term, behavior and environmental change generally take more time.

Evaluation must be valid and objective. If it is valid, it measures exactly what you want it to measure. If it is objective, it will not be affected by bias. Bias occurs when the attitude or expectation of the evaluator (or person being evaluated) influences the data being collected.

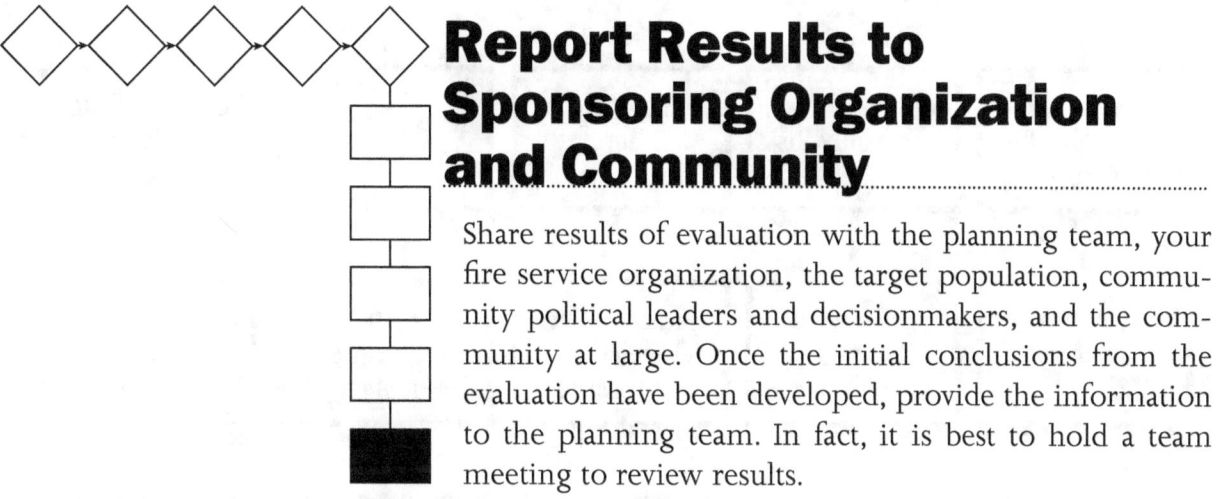

Report Results to Sponsoring Organization and Community

Share results of evaluation with the planning team, your fire service organization, the target population, community political leaders and decisionmakers, and the community at large. Once the initial conclusions from the evaluation have been developed, provide the information to the planning team. In fact, it is best to hold a team meeting to review results.

If a revision of the program is desired, it is important that everyone involved in conducting the program understands the reasons for changes.

Once the program is over, it is time to report the results of the program to the target audience, the organization, and the community. The conclusions from the evaluation process are the basis for this report.

Summary

Some essential steps must be taken to create a successful evaluation. First, make a commitment to do the evaluation well. Realize that the process will take time and effort to produce tangible results. Then, seek help. Ask someone who performs evaluation on a regular basis to help you get started.

Work hard to ensure all evaluation is done in a valid and objective manner. Keep an open mind and be prepared to make adjustments according to the findings of your monitoring and evaluation efforts.

Conclusion

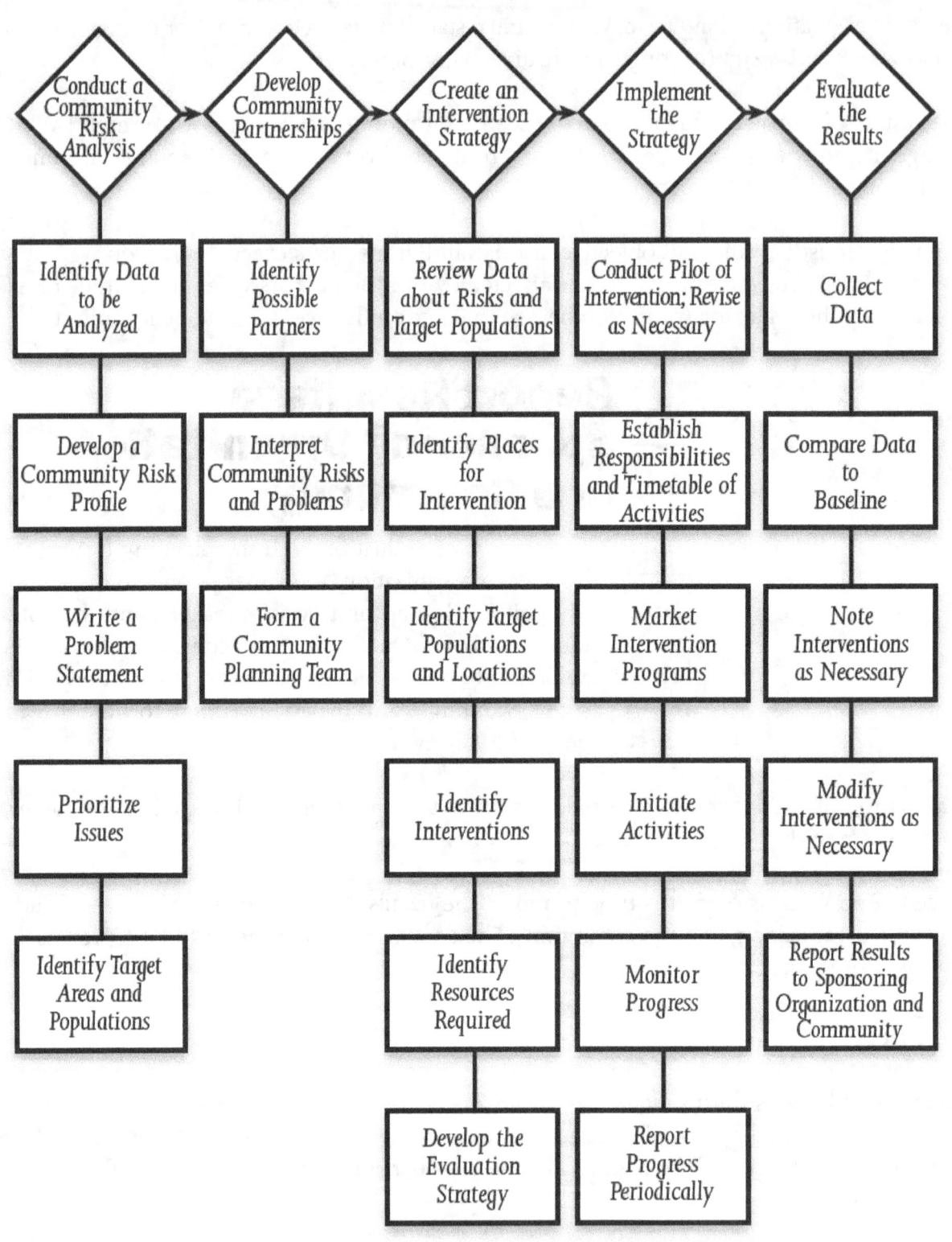

Conclusion

Community educators commonly have the misconception that once a program is implemented the work is over. Nothing could be further from the truth. The linear five-step planning process illustrated here represents the beginning and the end.

The education process never ends

The community education process is a never-ending cycle of analysis and revision. It is a series of five-step planning processes that overlap and continue through time.

Why is this a continual process? Communities and cultures change over time. The fire and injury problems facing communities today are different from those of 30 or 20 or even 10 years ago, in some cases. This community change process always will continue. Our educational programs also must reflect these changes.

To keep up with—and be ahead of—changing community problems, the educator must "close the loop" on the five-step planning process by connecting the evaluation step with the analysis step.

The information gathered through evaluation—the good and the bad about the program results—is the information that is used to begin another community analysis. This is a continual cycle that is designed to ensure that you and your community team are always aware of the fire and injury problems in the community. From this information current programs can be revised to meet the changes. In some cases new programs can be designed. In addition, the information can be used to help plan budgets and forecast future educational and operational needs.

Restarting the planning cycle

How often should the planning process be repeated? That question can be answered only by each individual community. Smaller communities that have fewer residents, neighborhoods, and socioeconomic/sociocultural groups will see a slower change in fire and injury problems and may require a community analysis only every 5 to 8 years. Larger communities composed of numerous cultures and groups will change at a faster pace and may require a community analysis every 2 or 3 years. In any event, it is up to the community team to determine how often a planning process is needed and then to see that the process is implemented.

However, this doesn't mean that a constant eye shouldn't be on fire and injury statistics. Regularly analyze available data on causes of fires, location of fires in the community, changing demographics and economic conditions, and types of injuries. Reviewing these data leads community educators to be proactive instead of reactive to community problems. In other words, watching for trends and patterns can help you identify problems early and put a program in place prior to the problem escalating. This should be the objective.

It all begins with you

Sound impossible to achieve? It may be at first. But after the first planning process is complete it will be easier to believe. The results of the process can be amazing. Hundreds of communities all over the country have completed successful planning processes and have thereby reduced the number of fires and injuries in the community. Any community can be successful in reducing fires and injuries through the use of the five-step planning process. It all begins with you.

Bibliography

U.S. Fire Administration. *America Burning: Recommissioned.* Washington: Author, 2000.

National Commission on Fire Prevention and Control. *America Burning.* Washington: U.S. Fire Administration, 1973.

National Safe Kids Campaign. *Childhood Injury.* Washington: Author, 2000.

_____. *Motor Vehicle Occupant Injury.* Washington: Author, 2000.

_____. *Residential Fire Injury.* Washington: Author, 2000.

_____. *Rural Injury.* Washington: Author, 2000.

National Center for Chronic Disease Prevention and Health Promotion. *Youth Risk Behavior Trends: 1991—1999.* Atlanta: Centers for Disease Control and Prevention, 2000.

_____. *Fire in the United States: 1987-1996.* 11th ed. Washington: Author.

Kirtley, Ed. *Risk Watch: A Leaders Guide to Preventing Childhood Injury in the Community.* Quincy: National Fire Protection Association, 1998.

International Fire Service Training Association. *Fire and Life Safety Educator.* 2nd ed. Stillwater: Author.

Turnock, Bernard J. *Public Health: What It Is and How It Works.* Gaithersburg: Aspen Publishers, 1997.

Appendix
Sources of Additional Training and Information

Noted below are several sources that you may find helpful as you develop a comprehensive community education program for your fire service organization. Federal Emergency Management Agency (FEMA) publications are available through the FEMA Web site at www.fema. gov

Readings

1. *Self-Study Course for Community Educators* (National Fire Academy).

2. *Short Guide to Evaluating Local Public Education Programs* (U. S. Fire Administration).

3. *Solutions 2000: Advocating Shared Responsibilities for Improved Fire Protection.*

4. *Fire Risk Series* (U.S. Fire Administration).

5. *Reaching High Risk Groups: The Community-Based Fire Safety Program* (Rossomando & Associates).

6. *Reaching the Hard to Reach* (TriData).

7. *Proving Public Fire Education Works* (TriData and U. S. Fire Administration).

8. *Community Tool Box: An Internet Resource* (University of Kansas).

Training

The following National Fire Academy courses support the information provided in this guide:

1. *Community Education Leadership.*
2. *Discovering the Road to High Risk Audiences.*
3. *Presenting Effective Public Education Programs.*
4. *Developing Fire and Life Safety Strategies.*
5. *Strategic Analysis of Community Risk Reduction.*
6. *Leading Community Fire Prevention.*